Cell Phone Science

What Happens When You Call and Why

MICHELE SEQUEIRA

MICHAEL WESTPHAL

University of New Mexico Press | Albuquerque

Barbara Guth Worlds of Wonder

Science Series for Young Readers

Advisory Editors: David Holtby and Karen Taschek

Please see page 168 for more information about the series.

Printed in China by Four Colour Print Group
Production location: Guangdong, China | Date of Production: 10/15/2010 | Cohort: Batch I

14 13 12 11 10   1 2 3 4 5

LIBRARY OF CONGRESS CATALOGING-IN-PUBLICATION DATA
Sequeira, Michele, 1967-
Cell phone science : what happens when you call and why / Michele Sequeira, Michael Westphal.
p. cm. — (Barbara Guth worlds of wonder science series for young readers)
Includes index.
ISBN 978-0-8263-4968-2 (cloth : alk. paper)
1. Cell phones—Juvenile literature. 2. Wireless communication systems—Juvenile literature. 3.
Mobile communication systems—Juvenile literature. I. Westphal, Michael, 1968- II. Title.
TK5103.2.S47 2010
621.3845'6—dc22
2010028600

With special thanks to the University of New Mexico Center for Regional Studies.
Illustrations provided by Cedra Wood.

This book is for Annaka and David who constantly question and challenge us (in a good way!)

Contents

Acknowledgments   1

A Note from the Authors   3

Chapter 1: Introduction: What Is a Cell Phone?   5

Chapter 2: Plastic, the Shell and Face of Your Phone   17

Chapter 3: Cell Phones Don't Think Like You Do   31

Chapter 4: Semiconductor Chips, the Brains   49

Chapter 5: A Cell Phone's Logic Is Written in Its Software   73

Chapter 6: Getting the Message Through: Call Transmission   91

Chapter 7: Batteries Are Gas Tanks Full of Electrons   109

Chapter 8: Good Design: Bringing the Technologies Together   125

Chapter 9: Thank You for Turning OFF Your Cell Phone   135

If You Want to Know More About Cell Phones . . .   143

Glossary   145

Illustration Credits   166

Index   169

Acknowledgments

Writing a book is an enormous undertaking that no one (of sound mind, anyway) attempts alone—there is always a team behind the endeavor. We are no different.

We would like to thank *our* team for all the editing, images, artwork, scientific fact checking, suggestions for improvement, insights (like 10-year-olds may not know what a landline is!) and general guidance they gave us in writing this, our first book.

Thanks to:
Karen Taschek, our editor, for initially seeing talent where we didn't know it existed. Karen gave us a lot of writing guidance, coaching, and encouragement.

Alan Steele, Eric Soderberg, and *Alice Ribbens,* for reading our manuscript (sometimes more than once) for accuracy and giving us great feedback.

Billy Cottle, for lending us his phone store and the cell phone props for us to take pictures for the book.

Soren Loree, for lending us his cars and car models for us to take pictures for the book.

Clark Whitehorn, Kathy Sparkes, and *David Holtby,* who guided us in the publishing process.

This is the first book that we've *written*—we've *read* lots of others!

Everyone who agreed to model for our pictures: *John Anuske-wicz, David Bentley, Billy Cottle, Rebecca Dawson, Donald "Woo-gie" Gadomski, Rebecca King, Katie Marshall, Niko Stamer, Maura Talley, Grace Walker, Noah Walker, Annaka Westphal, and David Westphal.*

The *Telephone Museum of New Mexico* in Albuquerque and the *Bradbury Science Museum in Los Alamos,* for helping us with some photographs.

Your inputs and time have been greatly appreciated!

> All models are wrong . . . Some are useful.
> —*George Box, statistician*

A Note from the Authors

We wrote this book to help you understand the science of cell phones, which can be quite complex. We use models (analogies) to explain the concepts in this book.

But remember, models are simplified explanations of the real thing. So, in a way, they are "wrong." They don't—and can't—describe *everything* about a concept. But models are useful because they lead you to important insights and understanding: *models help in predicting behavior and outcomes, and that's why we use them.*

Because models are "wrong," all of them break down at some point. The key to using a model is to know under what conditions the model you're using breaks down and where, when, and how a new model better describes the concept or perception you're trying to understand. This can be tricky—when do you abandon your current thinking and use a different model? Sometimes that new model you're looking for may not even exist yet; it needs to be discovered.

This process of thinking, imagining, and discovering is the fun of science. We hope to encourage it in you. Enjoy!

Think of a toy model of an older car: you can't actually unlock the door with a key or crank down the windows. These aspects of the car don't work, even though a very lifelike model may include details like little metal keyholes and little glass windows in the doors. Still, the model car serves a useful purpose if you wanted to try to describe what an older car was like to someone who had never seen one.

Sir Isaac Newton is often credited with the idea of breaking down a problem into a simpler one—a model—and solving it, then checking to see if the solution for the model applies to the actual problem. This problem-solving technique is used in many fields, not just in science.

CHAPTER 1

Introduction:
What Is a Cell Phone?

The cell phone is an amazing communication tool. No one person invented it. It's the combination of a variety of technologies that have been developed over the last 200 years.

Dude! It was great—I scored a hatrick and the other team didn't know what . . . Hey, the driver just told us we're not getting there until late . . . Yeah, really late . . .

Mom, the bus driver said we won't get to camp until late tonight . . . She said we have a flat tire. Someone is coming to fix it.

I know, Carlene, and then he said . . . Oh, wait a minute . . . I won't get there until really late! The bus got a flat . . .

You what? Got a flat? OK, we'll get someone out there to help you right away!

EVERYONE KNOWS WHY WE LOVE OUR CELL PHONES

—We can talk to our friends anytime, anywhere—but how do they work? Many technologies had to be combined to get quick, cheap calls from the bus to your house or from one state to another. Before we tear off the flashy cover of your cell phone to see what's inside, let's look at how long-distance communication got where it is.

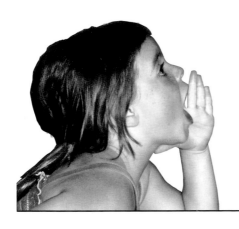

How Did We Get Along Before Phones?

WHAT DO YOU DO WHEN YOU WANT TO TELL SOMEONE SOMETHING BUT HE OR SHE IS TOO FAR AWAY?

You take out your phone, of course, but what did your ancient ancestors do? For many centuries, shouting, flag waving, and smoke signals were common ways to communicate. (Think about how much trouble that was. You're probably already glad for your cellie.) With fires or flags or shouting, the receiver gets the message at about the same time it's sent, but you can probably already think of drawbacks. It's hard to be clear:

someone waving or a puff of smoke can mean a lot of things, so you have to learn and use a code. And if you've got a lot to say, good luck—you'll be at it all day. Distance of communication is limited too. So, to send a complex, clear, or longer-distance message, a runner (messenger) might be better. "Snail mail," which grew

out of people sending written messages, is still popular. Mail has the added benefit of increased privacy. Everyone close enough to see the smoke can try to read your love note in the sky, whereas if you trust the messenger, only the receiver gets the paper love note. A drawback to messengers or mail is the time it takes to send and then receive a reply. And how practical is it to hire someone to run from the bus to your house to tell Mom something?

THE TELEGRAPH WAS A BREAKTHROUGH

Invented in the early 1800s, the telegraph can be considered the great-great-grand-uncle of text messaging. It uses Morse code: dots and dashes stand for each letter of the alphabet. The electrical pulses were transmitted over wires, where they became sounds or indentations in paper tape on the receiving end. Both the sender and the receiver of a message had to know the dot-and-dash Morse code, but for the first time, people could quickly send messages over miles.

This is a telegraph sounder. The person on the receiving end could quickly write down the code just by hearing it.

Samuel F. B. Morse, inventor of the telegraph, trained and successfully worked as a portrait painter. He lived in England after graduating from Yale University. During his return to America (by ship), he overheard a conversation on electromagnets that sparked his creativity—he invented the telegraph and the code with which to use it based on this overheard conversation!

This is a telegraph key (transmitter) made by Morse's company.

"Wire" Your Voice: from Telegraph to Telephone

Mr. Watson, come here!

Alexander Graham Bell.

DR. ALEXANDER GRAHAM BELL IS CREDITED WITH INVENTING THE TELEPHONE

He got it to work in 1876. Bell famously said, "Mr. Watson, come here!" When his assistant, Mr. Watson, who was in another room, heard him through the speaker on the phone, the golden age of phones began. Phones caught on quickly and were in wide use by the early 1900s. The telephone system, like the telegraph system, was a system of wires carrying electric signals that were encoded and decoded by the phone at each end (more about encoding and decoding in chapter 6). The big advantage the phone had over the telegraph was that the decoded signal reproduced the caller's voice, not a series of dots and dashes.

This is Bell's original telephone set-up.

HOW A LANDLINE (WIRE) PHONE WORKS IS BASICALLY THE SAME TODAY AS IT WAS IN BELL'S TIME

Your voice is electrically encoded, travels through wires, and is then decoded back into sound on the other person's phone. These days, far more calls to route mean that the system of getting the right phone to ring and connect is much more complicated than it was for Bell's first phone. And there have been some very clever improvements to allow multiple calls to travel over the same wires. The other great news about phones is that they are no longer black, huge, and clunky but are all the colors of the rainbow, tiny, and sleek.

An early commercial phone

(By the way, that cordless handset phone you use around your home works a little like your cell phone, but it communicates with your *landline* phone, not with a cell tower!)

THE HISTORIES OF THE CELL PHONE SYSTEM AND THE LANDLINE PHONE SYSTEM ARE VERY DIFFERENT

They each began differently and grew differently. So in addition to being different technologies, they have very different business and legal backgrounds. Landline phones were spread through the country by essentially one company. This gave that company a lot of power with people and the government. Cell phones have spread as a project of many different businesses that hope they can get people to buy their phones and services. Sometimes these businesses work together, but often, they are competing for the same customers.

One of these phones is not like the others . . .

That's right! The phones at right (the authors' phones!) are cell phones. The one's below are landline phones, or phones that are connected by wires to each other.

Dude! It was great . . .

Cell tower camouflaged as a tree

The Story Behind the Call

TO MAKE A CELL PHONE CALL, A WHOLE SYSTEM HAS TO WORK

This system is based on a wide range of science concepts and technology. And without the advances in technology that have made cell phones a joy to own, the big (and growing) list of tricks a cell phone can do wouldn't be possible. We're going to explore the science that makes your phone light, tough, cheap (inexpensive, not the bad kind of cheap) and effective. First, to really know how a cell call is made, let's look at the whole cellular system.

THE MOST IMPORTANT PART OF A CELL PHONE IS A RADIO

At their heart, cell phones are radio receivers and transmitters, like walkie-talkies (another technological great-great-ancestor of your phone). The signals that cell phones send and receive are radio signals that travel through the air—just like the signals that travel from one walkie-talkie to another—or the signals that travel from a radio station antenna to the antenna of a car radio.

CELL PHONES NEED CELLULAR ANTENNAE

Radio towers are often hundreds of feet tall. But you talk on a cell phone with people hundreds or thousands of miles away. And your phone's antenna is small—entirely hidden inside the cover. How can such a tiny phone transmit over such long distances? Well, it can't actually *transmit* very far, so it's a good thing that it doesn't have to. It only has to transmit to a nearby cellular antenna.

AN ENTIRE NETWORK OF ANTENNAE WAS BUILT TO ROUTE CALLS AS THE USE OF CELL PHONES EXPLODED

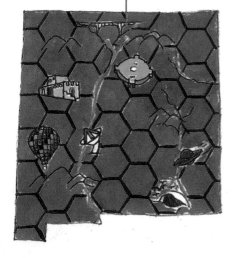

If you have cell service where you live and play, those antennae are nearby. Each antenna almost constantly communicates with all nearby phones (of the antenna's service provider) and with the rest of the antennae in the network. The word *cell* in *cell phone* actually refers to the way these antennae break up an area, called a *service area*, into *zones*, or cells, of radio reception. (So now you know why your cell phone is a cell phone and not a gizmo phone.)

TO START A CELL PHONE SYSTEM . . .

Somebody has to invest the money to design and build the parts and systems for cell phones. Keeping the cell phone system up and running has to make money for somebody or investing the money isn't worthwhile. If a cell phone company wants to keep making money (and they do), it has to keep its customers happy. This means things like being reliable in an emergency, being inexpensive, and also not trampling anyone's rights. If you can't immediately think of the brand names of several cellular providers, just watch a few TV commercials (not too many—you have better things to do with your time, like reading about how cell phones work). The relatively low prices we pay for cellular service are a result of competition between providers, as well as technical advances. The advertisements on TV and in other media are an unavoidable result of the same competition.

But, What Does the Phone in Your Hand Do?

A CELL PHONE IS AN AMAZING CREATION

- To power all of its electrical components (and there are many) and still stay portable, the cell phone uses a battery. (*Chapter 7: Batteries*)

- To be a handy communication tool, a cell phone must be able to send and receive signals from the cellular antennae. (*Chapter 6: Call Transmission*)

- To transmit signals, the phone has to talk to the cellular communication network, which itself must be able to send signals to the correct phone each time. (*Chapter 6: Call Transmission*)

- To do *this* well, the phone must be able to decode, or process, those signals—so that it knows when to ring, for example. (*Chapter 5: Software*)

- To process signals, the cell phone uses software. (*Chapter 5: Software*)

- To use the software, a cell phones uses semiconductor chips, which store and calculate information. (*Chapter 4: Semiconductor Chips*)

- Finally, all of the cell phone pieces have to fit into the attractive, light package that we can carry around. (*Chapter 2: Plastics*)

How does a cell phone do all of this? It's not magic, it's *science!*

Don't know what some of these pictures have to do with cell phones? Keep reading . . . you'll learn about the concepts these pictures represent!

Plastic, the Shell and Face of Your Phone

We start from the outside: your cell phone is encased in a strong, sturdy, lightweight, protective plastic shell. Plastic can be used in many different ways because it has properties unlike those of most other kinds of materials. Those properties can be selected to suit a variety of needs. In this chapter, we explore what these properties of plastics are, how they are controlled, and how they are used in your cell phone.

WHAT'S A PROPERTY, ANYHOW?

Plastic is so familiar to us that we scarcely recognize how much we come into contact with it. It's used to make drink bottles, eating utensils, car parts, school bus seat covers, bulletproof vests, jet fighter canopies, fibers for clothing, and of course, the outside of cell phones. What might you notice about different plastics: for instance, clear water bottles and dark green school bus seat covers? The water bottle is see-through, thin, smooth, flexible, lightweight,

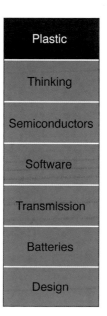

Plastic

Thinking

Semiconductors

Software

Transmission

Batteries

Design

and of course, waterproof. School bus seat covers are waterproof and flexible, but they aren't see-through, thin, or as smooth as a water bottle. When we describe the bottle as smooth or the seat cover as flexible, we're describing *properties* of the material.

WHAT PROPERTIES DOES A PLASTIC HAVE?

- It resists water (and many other chemicals).

- It's a good *insulator.*

- It's strong for its weight.

- It can be molded into complex parts, dyed different colors, or even covered with metal (to make those great-looking phone covers!).

- And it's relatively inexpensive to manufacture.

WHAT ABOUT CELL PHONES?

What properties does a cell phone shell need? First, think about what it does. It protects the delicate parts of the phone from rain, dirt, and the occasional drop on the sidewalk. The shell also needs to separate you and the electrical parts to prevent you from damaging the phone or getting a shock. Cell phone makers know that to sell the phone, its design must look good, so they use complex shapes, details, color, and shine to make the phone attractive and easy to carry and use. Plastic is a great choice of material for the shell because of its combination of properties.

WHAT IS PLASTIC?

How can a material like plastic have such a good combination of properties? It has to do with how it is put together. And to understand that, we must look closely at what it really is.

A scientist will tell you that what we typically call plastic really is an *amorphous polymer*. Don't worry about strange scientific words. We'll learn to say them and what they mean below. Scientists (and engineers) like to use very specific words

Different Kinds of Scientists

Put simply, science is the study of how things exist, interact, and change. Whether atoms or armadillos, Game Boys or galaxies, science looks at things in a systematic and detailed way. You may hear this described as the *scientific process* or *scientific method*.

Since the universe provides such a broad range of things to study, each scientist usually studies just one part of that range—the part that interests him or her the most. *Chemists* study chemistry, which is how atoms and molecules behave. *Physicists* study physics, which is how objects move and interact with each other. *Biologists* study biology, which is how living things grow and develop. *Materials scientists* study materials science, which is how the structure, properties, and processing of materials interact.

But just because you focus on one area doesn't mean you can't learn another. There's no reason why a chemist can't make important contributions to biology or a biologist to materials. In fact, some great discoveries have been made by scientists studying and working outside of their fields!

There are many more areas of science, and more are being created. Learn as much as you can about what interests you and you may find yourself a scientist.

Scientific Terms

To really understand the science and technology that go into making a cell phone, some words must be exactly defined. For example, the word *material* is a scientific term for *stuff*, as in the phrase: "the material from which things are made."

Exactly defining words is important. For instance, the phrase "Get some glasses" means one thing when your mom says it to you as she walks to the table with a pitcher of lemonade. It means an entirely different thing when a rude fan yells it at the home plate umpire. (And the word *glass* can mean an amorphous material when a scientist uses it!) We'll try to define what we mean as much as we can so that the scientific words we use won't be confusing.

for very exact meanings. But they also like to break concepts down into more and more simple ideas.

On the way to understanding plastics, we'll stack some blocks and balls, see how jellybeans and candy bars arrange themselves in a bowl, cook some spaghetti, and see how all of these relate to *molecules* and *atoms*.

WHAT DOES AMORPHOUS MEAN?

It actually has to do with how things are arranged inside a solid. There are two broad categories: *crystalline* (KRIS-til-in) and *amorphous* (uh-MOR-fus). More on what those "things" are on the next page.

CRYSTALLINE THINGS HAVE REGULAR ARRANGEMENTS

You can picture the meaning of *crystalline* by looking at the blocks below. Notice that you can see layers stacked on each other. In other words, the arrangement of the layers repeats.

The layers of these blocks are all alike, but in arrangements of other things, they can be different. In an arrangement of hockey pucks (or soup cans), for example, two different layer arrangements can alternate back and forth, repeating every second or even every third layer. Sometimes, you can even see layers in different directions. When scientists talk about *crystals*, they are referring to materials with crystalline arrangement of the atoms or molecules, not the fancy glassware from which you drink—the material *glass* is actually amorphous.

AMORPHOUS THINGS SHOW NO ORDER IN ARRANGEMENT

They don't really stack together like crystals. Instead of picturing a stack of oranges or soup cans (crystal), we need to imagine a plate of pasta or a bowl of candy bars.

In our plate of pasta, each piece is the same, just like each candy bar is the same. Sometimes, some pasta pieces look like they line up together, but it's only a few places here and there. Similarly, in our bowl of candy bars, a few might line up together, but that order only affects a very small area. There are no layers throughout the bowl, and there is no repeating pattern in the candy bars' arrangement.

WHAT IS A POLYMER?

We now know that there are some things in plastic that have a messy arrangement (like a bowl of spaghetti or jellybeans). The amorphous part of the term *amorphous polymer* tells us about the arrangement of the things. The polymer part tells us about what those things are. You may have already (correctly) guessed that they were atoms or molecules. Let's use a cube of sugar to explore how atoms and molecules relate to polymers.

A *substance* has the same properties throughout. This is different from a mixture, which is a combination of substances. Both can be solid, liquid, or gas

A MOLECULE IS THE SMALLEST PIECE OF A SUBSTANCE

It's easy to break a sugar cube into smaller lumps and to break those lumps into even smaller lumps. We eventually get powdered sugar (no surprise there). How much smaller do they get? You know that they become too small to see when dissolved in water. But if you drank some, it would taste sweet, so the sugar is still there. Scientists call the smallest piece of a material like table sugar a *molecule* (of that material). Sugar is a combination of *hydrogen*, *carbon*, and *oxygen*.

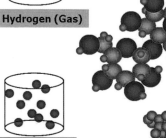

Hydrogen (Gas)

Oxygen (Gas)

Carbon (Solid)

A MOLECULE IS MADE UP OF ATOMS

A molecule is made up of even smaller parts. Materials like hydrogen, carbon, and oxygen aren't combinations of other materials—they are called *elements*. The smallest piece of an element is called an *atom*. A sugar molecule is a combination of carbon, hydrogen, and oxygen atoms. Every molecule of table sugar has the *same atoms, in the same arrangement.* Although a sugar molecule is made up of carbon, hydrogen, and oxygen, it's very different from any of these atoms. So if we break this molecule apart, even a little, it's no longer table sugar.

WHAT DOES SUGAR HAVE TO DO WITH THE PLASTIC SHELL OF MY PHONE?

The phone shell isn't made of sugar, of course, but seeing the 45 atoms in the sugar molecule helps in picturing even bigger molecules. The molecules that make up the cover of your phone are *huge* (by molecular standards) and consist of repeating parts. They are long strands, or chains, that have patterns of atoms within them. The repeating parts of the strands are called *monomers* and the entire strand is called a *polymer*.

Digestion is the process of chemically breaking food down so it can be used by the body. We'll talk about *chemical reactions* in chapter 7

It's Greek to Me . . .

Mono means "one"
Poly means "many"
Meros means "parts"

Hydrogen

Carbon

Is this really what atoms look like?

No one has ever seen an atom through a light microscope. But scientists have studied atoms through experiments and with other kinds of microscopes and these images are a good model.

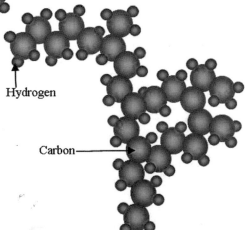

Hydrogen

Carbon

Everything is made up of atoms.

We can't see atoms, but they're there. They form almost everything, including the wind blowing on your face and the chair you're sitting on. We'll talk more about atoms and molecules in chapter 7.

Hydrocarbons

Maybe you've heard the term *hydrocarbon* because some pollution from our cars, power plants, and factories is called hydrocarbon emissions. Just as a plant growing in the wrong place gets labeled a weed, hydrocarbons are pollution if they're in the wrong place (like the air or the ocean). If you've seen the results of an oil spill, you know the huge damage that hydrocarbons can create.

Even biodiesel is a hydrocarbon. It's produced from animal or plant oils, like soybean oil, and can power diesel engines. But just like any other oil spill, a spill of biodiesel into a lake or stream would be pollution. It would harm living things (like ducks) near or in the water.

But when contained and used properly, hydrocarbons lead to wonderful things—including energy and plastics—for us.

SO HOW DO SCIENTISTS AND ENGINEERS MAKE STRINGS OF ATOMS INTO A CELL PHONE COVER?

Making the plastic (polymer) is kind of like cooking. It depends on ingredients: which ones, when, and how much of each are added. And like cooking, the processing (time, temperature, stirring, and so on) is important to the final result. The shape of the cell phone itself is typically created by melting polymer beads and then pushing the liquid into molds with pressure (injection molding). But just as pasta comes from eggs and flour, which come from chickens and wheat, there's more to the story. To create the shell takes just the right blend of ingredients and many careful steps.

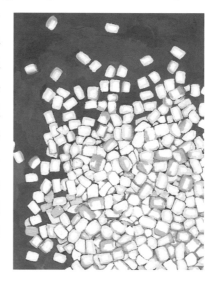

POLYMER BEADS THAT GO INTO THE INJECTION MOLDER START IN OIL FIELDS

The starting material for plastics is *hydrocarbons*—strands of carbon-and-hydrogen monomers. Almost all hydrocarbons used to make plastics start out as oil and natural gas that come out of the ground. They

can't be used directly, though—they must be processed first. *Refining* of hydrocarbons involves boiling and recondensing them.

DID YOU JUST SAY MY CELL PHONE IS MADE OF OIL?

Yes. The shell certainly is and so are a lot of the clothes, shoes, and cosmetics (yes, there's oil in that lipstick) that we see and use every day. Remember the list of plastic things at the beginning of this chapter? Bulletproof vests, water bottles, and bus seats all come from oil. There are some fabrics that are made from polymers refined from bamboo and other renewable sources, but they're a very small fraction of the polymer industry right now.

Refining is the process of making something more pure. Sometimes that means taking things out. Other times, it means causing a chemical reaction to create what you want. Different materials are refined in different ways.

DIFFERENT MONOMERS HAVE DIFFERENT PROPERTIES

A plastic that combines different monomers is called a *copolymer*. Its properties can be finely controlled by con-

trolling the type and amount of the different monomers that are used to make it. Chains can be formed from a repeating pattern of monomers, called *block copolymers,* or from a random pattern, called *random copolymers.* The addition of different elements can further affect the properties of the plastic.

THE ADDITION OF DIFFERENT ELEMENTS AFFECTS THE PROPERTIES OF THE RESULTING PLASTIC

Adding elements to the hydrocarbon chains changes the way the atoms attach to each other within the molecule and changes the way the molecules themselves interact. Some plastics, like *polystyrene* are made only of carbon and hydrogen. The carbon atoms attach to each other to form the strands and the hydrogen atoms attach to the carbon atoms. Other plastics contain other elements, such as oxygen, chlorine, fluorine, nitrogen, silicon, sulfur, and phosphorus.

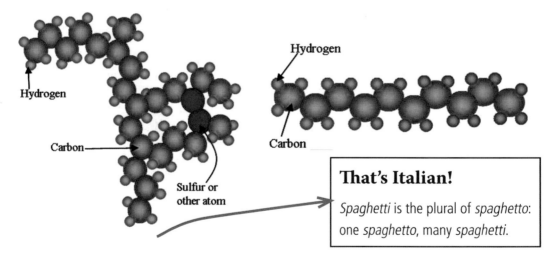

Hydrogen

Carbon

Hydrogen

Carbon

Sulfur or other atom

That's Italian!

Spaghetti is the plural of *spaghetto:* one *spaghetto,* many *spaghetti.*

SPAGHETTI CAN SHOW CROSSLINKING

To imagine how the structure of a plastic can be changed by the addition of different elements, think of a plate of spaghetti. A typical spaghetto is *linear.* If each strand is covered in a sauce with lots of olive oil, it can slide over others. If you then tried to pick up the spaghetti with your fingers, you would only get the few strands you actually held and the rest would slip back onto the plate. As a whole, the

glob of spaghetti doesn't have much strength: it can't even keep itself together! But if instead of adding sauce, the cooked spaghetti is left in the strainer and dried a little, the strands will stick to each other, probably at many different places along their length. This time if you picked up the spaghetti with your fingers, you would get almost all of it in a big clump. This glob of spaghetti, although unappetizing, has a lot of strength and can't be pulled apart easily.

The same behavior occurs in plastics: the polymer strands of carbon and hydrogen can "stick" to each other along their lengths. This is called *crosslinking*. Pasta cooks try to avoid crosslinking; polymer scientists and engineers try to control—and often encourage—it.

Spaghetti is a good model for imagining long strands of hydrocarbons. But not all polymer chains are linear—some are branched. All polymer chains can crosslink, though.

Plastic, the Shell and Face of Your Phone   27

CROSSLINKING IS VERY IMPORTANT TO INCREASE THE STRENGTH OF A PLASTIC

Crosslinking keeps the polymer strands from moving over each other, like our sticky spaghetti. Crosslinking sometimes occurs when the hydrogen atoms attach to each other without detaching from the carbon atoms. Polymer engineers can get more crosslinking by adding different elements that replace some of the hydrogen or carbon in the carbon-and-hydrogen monomers. These elements then attach to each other, bonding—gluing—the strands of polymer at various places along its length.

These are called *hydrogen bonds*

THE AMOUNT OF CROSSLINKING IN A POLYMER CONTROLS HOW RIGID THE MATERIAL IS

A polymer in which only one in 100 molecules (carbon-and-hydrogen strands) is crosslinked is rubbery and is called an *elastomer*. It has *memory*, which means that if you stretch it, the strands move over each other but then bounce back into place. You may have heard of *vulcanization*, in which rubber (an elastomer) is made more rigid by increasing crosslinking. Vulcanization is used to make car tires strong and hockey pucks hard.

CROSSLINKING IS WHY PLASTIC IS THE IDEAL CHOICE FOR THE SHELL OF YOUR CELL PHONE

The polymers known as plastics have about three times as many crosslinked molecules (about one in 30) as elastomers. This high amount of crosslinking is important. Plastics are more rigid than elastomers, and this rigidity is ideal for protecting the innards of your cell phone: the phone shell doesn't bend and crush the delicate electronics when you drop it. At the same time, the high amount of crosslinking makes the plastic lightweight because it creates space between the carbon-and-hydrogen strands. Metals, which are also strong, are heavier since they don't have a lot of space in their structures. Plastics can be rolled into sheets or shaped in molds to make complex parts; the crosslinking helps here, too, in keeping the shape of the molded plastic.

Most metals are more *dense* than plastics. Density is the amount a substance weighs (on Earth) divided by its volume.

CHAPTER 3

Cell Phones Don't Think Like You Do

Let's look inside your phone. Your cell phone's "brain" is actually a computer: a set of tiny pieces of processed semiconductor, or chips, all working together. On the most basic level, each chip does only a few things, but it works very well and blindingly fast to accomplish complex tasks. In this chapter, you'll start to see what a computer chip does and why this is important for cell phones.

Abstraction

THINKING VERSUS COMPUTER PROCESSING

Look carefully at the figure on the next page—do you see a word? Your eyes and brain see the word even though the display is actually an arrangement of shaded and empty boxes. You know it's just boxes, but you gather meaning from the pattern. If you were asked to write the word, you would draw a line, some curves, a circle, and so on. You probably wouldn't even think about the steps needed to write *Bob* since you are so familiar with not only the letters but the word itself.

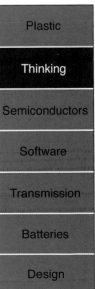

Plastic

Thinking

Semiconductors

Software

Transmission

Batteries

Design

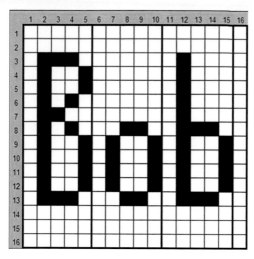

OUR BRAINS ARE WONDERFUL AT ABSTRACTION

That is, the brain can take in a few, important details and then fill in the rest. Every now and then, an optical illusion reminds us that this isn't a perfect system, but usually it works very well. The ability to *abstract* helps us see and react to our world: we recognize faces, we run from angry lions, and we know not to go through a door marked *DANGER: Do Not Enter.* In the drawing above, you abstracted meaning from the shaded and empty boxes. Computers don't abstract the way that you do. Everything they do must be taught to them at a very basic level.

HOW FAST CAN YOU COUNT?

Can a computer do something that the people who built and programmed it don't know how to do? This is a tricky question. Even the simple computer that helps a car engine run makes many adjustments every second. No human is quick enough to match its speed. But that or any computer has to be taught how to do every task. One way to think about—or model—what a computer does is to consider it starting with two simple abilities: keeping track of numbers and counting. We'll see later that even those abilities are designed into the computer, or taught. The computer never sees letters the way you did. So, to instruct a computer to display a word (or to display a text message or to tell you that you have an incoming call) requires that humans use a different way of thinking. We have to start very simply—for example, with tracking and counting. Building on this simplicity pays off because with the right instructions, a computer can keep track of vast quantities of numbers and count very fast—far, far better than a human can.

LET'S TEACH A COMPUTER TO DISPLAY THE WORD BOB

We know the computer must use pixels like the shaded boxes. How should it go through the set of 256 pixels in the 16-row-by-16-column (16 x 16) grid?

- Does it go across the first row from left to right, then start from left to right on the second row?

- Does it go down the first column, then come up on the second column and then down the third?

When instructing a computer, simple directions like these are important to explain—otherwise, the computer won't know what to do.

When we say computer, we mean a set of processed semiconductor chips (and the hardware around them) that work together. Computers track satellites orbiting around the earth to help you know where you are, they count the deposits made in each bank every day, and they are used to measure how much medication a surgery patient receives after a medical procedure—among a whole lot of other uses.

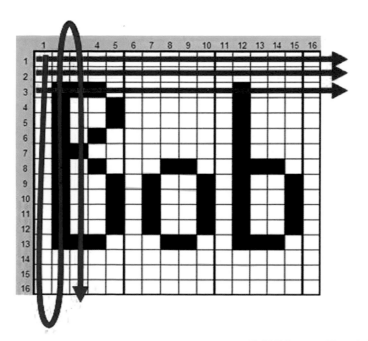

A *pixel* is a small, distinct element that, when used with other pixels, creates an image.

IF WE TELL THE COMPUTER TO . . .

Go across the rows from left to right and when it gets to the 16th square in the row, start on the next row on the left (column 1) and go left to right again. We might also tell it that an unshaded pixel is represented as 0 and a shaded pixel is represented as 1 and each row is a separate line. Then the first six rows in the display would be read as:

0000000000000000	0100100000010000
0000000000000000	0100100000010000
0111000000010000	0101000000010000

IF WE CHANGE THE INSTRUCTIONS . . .

So that we go down the odd-numbered columns and up the even numbered ones, then the computer would read the display differently. The first six columns would be read as:

0000000000000000	0001000010100100
0001111111111100	0001100011110000
0010001000001000	0000000000000000

TWO REPRESENTATIONS FOR THE SAME WORD

These two representations of the word each have a total of 256 digits (16 rows x 16 columns), but the sequence of ones and zeroes is very different. We told the computer that the 19th digit in the first sequence corresponds to the 2nd row, 3rd column, and that in the second sequence it corresponds to the 14th row, 2nd column. Knowing the order in which to display the sequence, as well as the sequence itself, is critical.

ONES AND ZEROES ARE HARD FOR A HUMAN TO READ, MUCH LESS UNDERSTAND

It's a lot of effort to keep track of where we are in a sequence of ones and zeroes. It's easy to get lost or make a mistake. But for a computer, it's easy to interpret these ones and zeroes very quickly.

THE BIGGER THE MESSAGE, THE MORE ONES AND ZEROES

The display in our example is a small, simple one. It uses only 256 pixels and displays a single word. If a computer is to display a longer word or many words (like an e-mail or text message), it's easy to see that a much bigger sequence of ones and zeroes will be necessary.

Reuse

ANOTHER LEVEL IN TEACHING A COMPUTER

Let's build another level of learning for our computer now that we see a way to teach it to display a word using simple instructions. Suppose we want to display a different word, *cob*. We could create a whole new 256-digit sequence, but we know that much of that sequence would be repeated since the last two letters of *Bob* are the same as those in *cob*.

A SEQUENCE FOR EVERY LETTER

Instead of creating a separate 256-digit sequence for this new word, we could instruct our computer to divide the grid into three smaller grids and a single, blank column. Each letter would use a five-column-by-16-row grid (white, yellow, and purple), represented by an 80-digit sequence, and the end of the word would be represented by the single blank column (green). With our new system of organization, we can *reuse* part of the old word for the new word instead of having to re-create a 256-digit sequence.

STORING 52 SEQUENCES IS EASIER THAN STORING 5,000

This reuse of sequences already used can help us to save time and memory space: instead of writing 256-digit (or more) sequences every time, we can teach the computer that one 80-digit sequence is the display for *o* and a different 80-digit sequence is the display for *b* and so on. The computer needs to keep track of only 52 of these 80-digit sequences to display every letter of the alphabet (upper and lowercase). This in turn allows it to display any word in the English language without having to have the space to *store* each one.

ARE WE READY TO TEXT?

If the computer has the memory to store the ones and zeroes, we now have a way to create words from simple instructions. It's a very basic step, but we're going in the right direction. Before we can text, we have to add many more abilities.

We can reuse the 80-digit sequences for o (yellow) and *b* (purple) and use the 16-digit sequence for a blank column (green) to indicate the end of a word.

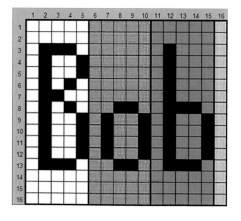

Only the white sequence needs to change for the new word!

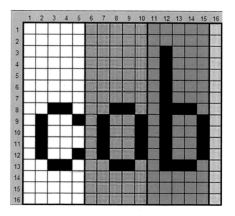

Building Complexity

TEACHING SIMPLE DECISIONS

By connecting individual simple decisions, we can program a computer to make complex decisions. Decisions may sound like things that take judgment (a human quality), but the kind of decisions a computer makes are much more boring. They are based on simple inputs: combinations of ones and zeroes to be exact.

THE LIGHT SWITCH GOING ON AND OFF IN YOUR REFRIGERATOR IS AN EXAMPLE OF A DECISION

Remember when you thought that a gnome in your refrigerator turned the light on when the door was opened? No matter how fast you opened that fridge door, you never quite caught a glimpse of him, did you? By now you know that it's not a gnome but a spring-loaded switch that turns on when the door opens and off when the door closes. This switching on and off is a simple decision. It might not seem like much—and it isn't—but something changes: the *state* your light was in (off or on) changed. What's more, you know how to control it. The position of the switch controls the light.

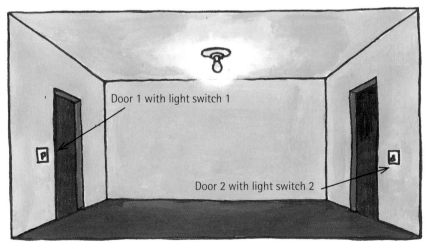

Door 1 with light switch 1

Door 2 with light switch 2

What is a logic gate?

It's a description of a set of conditions and the outcome for each. Each light switch described here is an example of a logic gate.

A DOUBLE LIGHT SWITCH SHOWS SIMPLE LOGIC

Let's get more complex. If you explore your house, apartment, or school, you'll find a room in which the lights can be switched on or off from two different places, usually by different entrances. Each of these switches is a NOT logic gate for the light. This kind of logic is useful here because it allows either switch to make the light do the opposite of what it was doing before. If you enter the darkened room from one entrance and flip the switch to turn the light on, it stays on. But you don't need to go back to that same switch to turn the light off: you can leave the room through the other door and flip the switch there to switch the light off. At each door, the switch changes the state of the light.

NOT Logic Gate for the Light Switch

If the light is . . .	Then turn it to . . .
ON	OFF
OFF	ON

OTHER LOGIC GATES: AND AND OR

Now let's look at some other decisions. The NOT gate took one input into account. AND and OR gates take two inputs into account and give different results. The example below will help to explain.

IMAGINE . . . YOU ARE IN CHARGE OF OPERATING A RAILROAD THAT CONNECTS TWO VILLAGES SEPARATED BY A MOUNTAIN

You have two sets of track. The "Alrededor Line" goes around the mountain. It's long and slow but very reliable. The "Por Line" goes through the mountain using a bridge and a tunnel. It's much faster unless there is a problem with the bridge or tunnel. You have one signal station at the bridge and another station at the tunnel to report track conditions so you know when each is safe to use. At each signal station, your faithful observer watches the track for problems. In each signal house there is a light switch that can be turned on, indicating a safe track. Each switch is connected by wires back to a light in your depot. There is a light for the bridge and another light for the tunnel. If a heavy rain washes out the bridge or a rock slide blocks the tunnel, you receive the signal and use the Alrededor Line. From the perspective of the Por Line, the bridge *AND* the tunnel have to be safe for the line to be safe.

Why are logic gates called gates?

No, it's not because of Bill Gates, co-founder of Microsoft. A gate is something that controls something else. So, you can think of a logic gate as something that controls the decision.

CREATING LOGIC GATES

Now, to train your new assistants at the station house so that they can make the decisions about signals without you, you decide to write down all the possible signals and outcomes for the Por Line. You soon see that there are four of them, only one of which has the outcome of the track line being safe.

Outcomes for the Por Line:

1. Bridge safe (1), tunnel safe (1) » line safe (1)

2. Bridge not safe (0), tunnel safe (1) » line not safe (0)

3. Bridge safe (1), tunnel not safe (0) » line not safe (0)

4. Bridge not safe (0), tunnel not safe (0) » line not safe (0)

We can write these conditions in a decision table like this:

	Tunnel Safe (1)	Tunnel Not Safe (0)
Bridge Safe (1)	{1 1} » (1) Line Safe	{1 0} » (0) Line NOT Safe
Bridge Not Safe (0)	{0 1} » (0) Line NOT Safe	{0 0} » (0) Line NOT Safe

Notice that we're also using numbers to indicate the condition of the track. For clarity here, we will use parentheses () to show a single input condition or outcome and braces {} to show more than one input condition. Using this notation, it's easy to use outcomes as input conditions. This will be important when we combine decisions (below)!

When we write just the numbers, the table now looks like this:

AND Gate

Condition X (Bridge)	Condition Y (Tunnel)	Outcome (Safe)
1	1	1
1	0	0
0	0	0
0	0	0

CONGRATULATIONS!
You have written one of the simple decision sets that a computer uses: the AND gate. You can think of it like a set of instructions, because that's what you're doing: you're telling someone (or something) what to do based on the inputs.

A LOGIC GATE IS A SIMPLE WAY TO DESCRIBE THE INPUTS AND OUTCOMES OF A DECISION TABLE
Notice that the outcome in the AND gate is 1 *only if* both conditions are 1. That is, if condition X *and* condition Y are 1, then the outcome is 1. In the case of our track, condition X corresponds to the bridge, condition Y corresponds to the tunnel, and the outcome tells whether the Por line is safe to use or not.

Instead of training an assistant, you could use a very simple computer to indicate which track to use. The "computer" in this case is just a logic gate. Each input and outcome is simplified to a 1 or a 0. The decisions are shown in the AND gate above. Some other basic logic gates are OR and NOT.

This table shows an OR gate, in which the outcome is 1 as long as one of the conditions (or inputs) is 1.

OR Gate

Input X	Input Y	Outcome
1	1	1
1	0	1
0	1	1
0	0	0

Remember our NOT gate? It has only one input, and the output is the opposite of the input. Here it is using only numbers.

NOT Gate

Input	Outcome
1	0
0	1

LOGIC GATES DESCRIBE A PATTERN FOR A DECISION

The basic idea behind how computer scientists and engineers build complexity into a computer is the connection of logic gates. The output of one can become the input to another. By linking logic gates together, it's possible to describe a large number of possibilities, not just two or four!

LET'S USE OUR TRAIN EXAMPLE ONE MORE TIME TO DEMONSTRATE HOW COMBINING LOGIC GATES CAN DESCRIBE COMPLEXITY

We know that if the train isn't able to go over the bridge and through the tunnel, then it must go around via the Alrededor Line. Since the Alrededor Line is entirely on solid ground, we can use longer, heavier trains on it. We want to get as much cargo transported through our railroad as possible and as quickly as possible. So, this means that every time we can't use the Por Line, we want to load up and send our biggest, heaviest trains on the Alrededor Line.

In words, here's what our decision looks like:

Outcome 1 is the outcome for the Por line (our previous AND gate).

Outcome 2—what kind of trains to use—depends on what outcome 1 is.

Outcome 2:

If Por Line (outcome 1 = 1), then use the shorter train (0).

If Alrededor Line (outcome 1 = 0), then use the longer train (1).

Here are the logic gates—combined—to describe this decision:

AND Gate   NOT Gate

Condition X	Condition Y	Outcome 1	Outcome 2
1	1	1	0
1	0	0	0
0	1	0	1
0	0	0	1

So, now we can describe not only which route to take, but also which train to use!

LOGIC GATES CAN DESCRIBE ANY SCENARIO

That's the beauty of them. By using just ones and zeroes, it's possible to describe any situation. Here's another example:

Suppose you are a dispatcher for a rural emergency call center and your job is to get help to an injured person as fast as possible. You must decide whether to send the helicopter or the truck. The helicopter is much faster, but you can't send it into high winds or an electrical storm. You must send the truck under those conditions.

We will come back to these ideas in chapter 5.

Hardware and Software

BUILDING A COMPUTER FROM SWITCHES

Now we know that something as simple as two light switches in a row can be used as a basic building block of a computer. Should we try to build a computer that uses switches for input, a series of logic gates to process, and lightbulbs for output? Early computer engineers actually did this, much like our train example, only a few hundred times more complex. Even though we would consider these computers primitive, they were complex enough that the programmers had to keep track of what they were doing.

John Von Neumann (1903–1957) (top) and Alan Turing (1912–1954) (bottom), mathematicians who strongly influenced the development of the computer as we know it today.

(right): This person is loading punchcards into the computer, which takes up an entire room! Punchcards were what early programmers used to program computers. Obviously, troubleshooting your program was much more difficult in those days.

SEPARATING THE SWITCHES FROM THE INSTRUCTIONS MAKES PROGRAMMING EASIER

Logic is the decision-making ability that comes from inputs, logic gates, number tracking, and counting working together. As computers and their abilities became more complex, people started talking about better ways to use them. A revolutionary solution was suggested by mathematicians Alan Turing and John von Neumann in the 1940s: separate the computer—*hardware*—from the instructions—*software*. In other words, put some of the logic into how the

Courtesy of the Computer History Museum.

electronic components are wired together (the *circuits*) and put the rest of the logic into how they are *programmed* (more about these in chapters 4 and 5).

At the time, most computer circuits were custom-built to perform one task very well, but then those circuits required a lot of rebuilding to perform a different task. Accordingly, the computer programs would have to change a lot too. Turing and von Neumann's solution would require that the computer circuits (hardware) be built so that they could be used over and over, with different instructions. At the same time, the instructions (software) could be used on different computers with minor adjustments. Lego bricks can help to show how separating instructions from hardware works.

LEGO BRICKS CAN BE REUSED

Lego bricks can be used to create a variety of things, from statues of Harry Potter characters to scaled models of cities. Each brick can connect to almost any other, *but only in certain ways.* The bricks must connect so that the bottom of one fits onto the top of another and so that they are always either aligned or perpendicular to each other (no weird angles when they connect). In this way, the bricks follow rules. They have logic. You can take apart a structure and build something else out of the bricks. With Legos, this ability to use the bricks in different ways is part of what makes them so great as a toy.

INSTRUCTIONS CAN BE REUSED

If you had a number of Lego bricks in front of you and a set of instructions that showed how to create a car, you could follow those instructions to create that car using those bricks. Good instructions also follow rules—they have logic too. They tell you what pieces you need and how to put them together.

But when you finish playing with the car, you could take it apart and create a robot using some of the same bricks and a different set of instructions. The bricks weren't used up—they still worked as bricks. And if you wanted to create the car again, all you'd need to do is take apart the robot and put the bricks together again using the car instructions.

LEGO BRICKS AND INSTRUCTIONS ARE INTERCHANGEABLE

What's more, you could photocopy your set of instructions for a friend and he could use his own set of Lego bricks to build a car or a robot just like yours. The instructions can be used with anyone's set of Lego bricks, and the Lego bricks can be used with anyone's set of instructions— each is separate from the other.

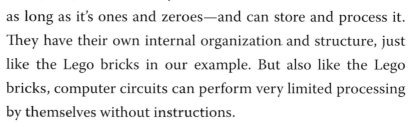

When we talk about hardware, we mean just the semiconductor chips and their components that do the processing and information handling, not the plastic shell, keypad, speaker, and similar parts. In this sense, many gadgets today, like toaster ovens and washing machines, have computers in them.

HARDWARE IS LIKE LEGO BRICKS

Computer circuits, or hardware, physically keep track of and count numbers. They use their multitude of *transistors* to do that (more about transistors in chapter 4). They can accept information (*data*) from a variety of sources— as long as it's ones and zeroes—and can store and process it. They have their own internal organization and structure, just like the Lego bricks in our example. But also like the Lego bricks, computer circuits can perform very limited processing by themselves without instructions.

SOFTWARE IS LIKE THE INSTRUCTIONS TO BUILD A CAR OR A ROBOT

Software uses the information (data) stored in the hardware to display a text message, to take a picture, or to re-create someone's voice in a phone call. It can only work on certain types of hardware, just like our car or robot instructions required a certain number and types of Lego bricks. And also like our Lego instructions, software can be used over and over again on any number of computers, but it's only useful on a computer (hardware). By itself, software can't do anything.

COMPUTERS AND SOFTWARE CAN BE INTERCHANGEABLE

The benefits of separating the hardware from the software of a computer are many. Computer circuits can be "taught" easily and quickly by simply inserting a disk or *downloading* a program.

The transistors on the chips can be made to operate faster—that is, the circuits can be manufactured to operate more quickly. And they can be made smaller and smaller so that they take up less room while delivering more processing capability.

Software can be reused many times and in many places—making the huge effort to create it worthwhile. It can be made more efficient with each revision. Your cell phone takes advantage of all of these benefits.

At the beginning of this chapter, we saw how the word *Bob* was shown using pixels. In that imaginary example, each letter in the word needed 80 pixels and each pixel was described by a digit: a 1 if the pixel was displayed black or a 0 if the pixel was displayed white. If each digit were actually a transistor switched on or off, then you'd need 80 transistors to store each letter pattern. So, imagine how many transistors you'd need to display just the letters in this box—and you can see why you'd need a lot of transistors in your computer to do much of anything.

CHAPTER 4

Semiconductor Chips, the Brains

The circuit board inside your cell phone (left) has many different chips, each packing a huge number of transistors. The transistors are used to process the huge number of 1s and 0s your cell phone uses. (Remember in chapter 3 that it took 256 pixels, each controlled by a transistor, to display *Bob*.) So, each transistor needs to be as small as possible to make your phone portable. Making something this tiny means we can't just use our hands to put it together.

Semiconductors Made Phones Portable . . . Eventually

THE FIRST MINICOMPUTERS WERE HUGE IN SIZE AND SMALL IN ABILITY

They took up entire air-conditioned rooms (in a time when air conditioning was rare!) and couldn't even do as much as the computer in your cell phone today. They were called *mini*computers because they were, in fact, much smaller than earlier computers. Those took up even more room, had even less capability to compute (*computing power*), and broke down more often. So the minicomputer represented a huge improvement.

Computing power is measured in *instructions per second.* We'll touch on this more in chapter 5.

Plastic

Thinking

Semiconductors

Software

Transmission

Batteries

Design

USE OF INTEGRATED CIRCUITS SHRANK
COMPUTERS' SWITCHES

Early computers relied on *vacuum tube* technology. The vacuum tubes used for the transistor switches were made of glass and contained wires that carried *electricity*—similar to a lightbulb. And, like a lightbulb, these transistors got very hot, which made them switch unreliably.

Now, imagine carrying a phone full of vacuum tubes! You'd need a lot of padding to protect the vacuum tubes from a drop . . . *if* you could lift the phone in the first place!

A transistor is an electronic device that can be used in different ways, but in computer circuits, they are almost always used as switches, like the switches for logic gates discussed in chapter 3. Minicomputers were among the first computers that were built using *integrated circuits (ICs)*. ICs shrank the transistors to unimaginably small sizes (for that time) and connected all of them together on a single *semiconductor chip*. Since these advances have had such a huge impact on the development of technology over the last few decades, we'll talk about transistors and integrated circuits a little more.

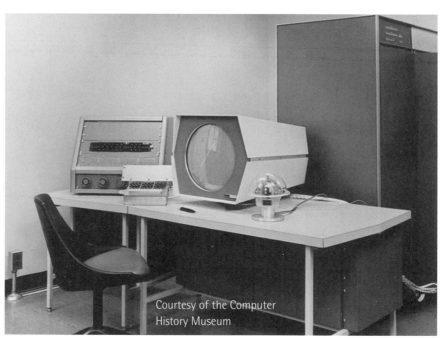

Courtesy of the Computer History Museum

This is a DEC PDP-1 minicomputer. (PDP stands for **Programmed Data Processor** and yes, that is an adult sized chair in the photo.)

The ICs are contained in the large box behind the desk, and the circular glass is the display tube. There is no keyboard or mouse—the keyboard wasn't needed and mice hadn't been invented yet. To program it, the trained programmer would press the buttons and flip the switches on the console to the left of the circular screen.

Do you like to play video games on your cell phone? A PDP-1 was used to invent and play the first video game. It was called Space Wars and was developed by college students in their spare time.

A TRANSISTOR ON A COMPUTER CHIP IS VERY SMALL

We saw in chapter 3 how quickly the need to store many 1s and 0s in computer circuits grew. So, very small transistors are critical for the computer in a cell phone (unless you want to lug around a heavy, bulky block that you have to recharge all the time). In fact, transistors many times smaller than the *diameter* of a human hair are needed. At sizes as small as these, we must look even more closely at the materials themselves to understand how these very small devices work—we must consider how the *atoms* in transistors work together.

ICs TODAY ARE MADE FROM CRYSTALLINE SILICON

The *silicon* transistors in ICs take advantage of an interesting property of pure silicon: it acts as an *insulator* at some *voltages* and as a *conductor* at other, higher voltages. So, silicon is part of a class of materials known as *semiconductors*, meaning "half-conductors."

A semiconductor acts like a rolled-up noisemaker that children (and some adults) blow at parties. When the noisemaker is just sitting

Crystals of other materials, most notably *gallium arsenide*, are also available but are not as widely used. Transistors created on gallium arsenide wafers are faster than transistors created on silicon wafers but are more expensive and not as easy to produce. We'll focus here on integrated circuits made from silicon.

on the table, it has air in it and around it, but the air is still. When you hold the noisemaker up to a fan or blow on it gently with low pressure, a little air moves, but the springy tube doesn't unroll and the air just sort of leaks away. When you actually blow hard, with pressure, the paper tube unrolls and the air flows out the other end fairly easily. When you stop blowing so hard, the paper tube pulls back and it no longer acts like a pipe.

Although the actual physical changes in semiconductors are nothing like those in a noisemaker, their behavior is a lot alike if you substitute electrons for the air and remember that air pressure is like voltage.

Lower-pressure air: there's air in the tube of the noisemaker, but the pressure is the same as that of the surrounding air.

Higher-pressure air: here the air in the tube has more pressure than the surroundings, making the tube inflate and allow the air to flow from one end to the other.

Atoms and Their Components

As we learned before, atoms are the smallest part of an element. An element, like carbon or hydrogen or oxygen, is a pure substance and not a combination of other elements, like sugar.

Each atom is made up of smaller particles. Physicists are still discovering and debating about all the different particles contained in atoms, but they and chemists agree that every atom does contain at least one *electron* and one *proton*. Each atom has an equal number of protons and electrons; some atoms contain an additional type of particle called a neutron.

Chemists distinguish between the different elements according to how many protons each atom of the element has. The *periodic table of the elements* shows all the elements discovered so far and organizes them by how many protons each has.

All the atoms of an element have the same number of protons (and so the same number of electrons), but some atoms of an element have different numbers of neutrons. Atoms with different numbers of neutrons are called *isotopes*.

All of the atoms drawn here are *neutral*; that is, they have no electric charge. More on charges and atoms in chapter 7.

It's the number of protons that make each element different. A proton in one atom is just like a proton in any another. So, when we talk about a proton in hydrogen, it's the same kind of particle as a proton in helium. The same is true for electrons and neutrons: it's the number of them that matters.

Helium Atom:
2 electrons, 2 protons.

This is an isotope of the other Helium atom—note the neutrons compared with just one in the other isotope.

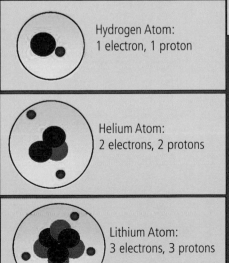

Hydrogen Atom:
1 electron, 1 proton

Helium Atom:
2 electrons, 2 protons

Lithium Atom:
3 electrons, 3 protons

Remember that 9X10^-2 is written as: 0.09

The difference between mass and *weight* can be confusing. Weight is how much gravity pulls on an object. So if you're in space, where there is almost no gravity, you may have no weight but would still have the same mass you have on Earth. If you're on Jupiter where gravity is stronger, your weight would be greater than here on Earth, but you'd still have the same mass.

Big Numbers, Small Numbers, and the Size of an Electron

People working in science often use very big and very small numbers to describe things. Very big numbers describe the distance to a star. Very small numbers are used in designing a new cell phone. As we talk about how electrons are used in transistors, think about how astonishingly light an electron is.

An electron has *mass* of about

$$9 \times 10^{-28} \text{ gram}$$

Which is:

0.0000000000000000000000000009 grams

But what does this mean? A simple way to think of mass is the amount of stuff that is in an object. The mass of a raisin is about 1 *gram*. So, to have the same mass as a raisin, you would need about one octillion electrons, which is written as

1,000,000,000,000,000,000,000,000,000

One octillion is such a big number that few people use it in everyday life. So what does it mean? Let's start with something familiar: blades of grass. Have you ever wondered how many blades of grass are on a soccer field? Let's say that there are about 3,000 blades of grass on every square foot (an area one foot long by one foot wide). That means that there are a little more than two million blades of grass on the entire soccer field. Two million is written as

2,000,000

Now, if we wanted to represent every electron in a gram with a blade of grass, we'd need a lot of soccer fields. In fact, we don't have enough room on the whole earth for them, even if we somehow covered the oceans with soccer fields! To have one octillion blades of grass, we would need about 66 million Earths, completely covered in blades of grass! Sixty-six million is written

66,000,000

This is still a huge number, which means that you would need an incredibly large number of electrons to equal the mass of a raisin.

Insulators, Conductors, and Material Structure

Electricity is the movement of small, *charged particles* like electrons.

An insulator is a material that doesn't allow electrons to move very well. Because of this property, they have many uses. The plastic cover around an electrical plug allows you to touch the cord without having electricity flow through your fingers or body, for example.

A conductor, by contrast, is a material that allows the fast movement of electrons. The tines—metal parts—in an electrical plug are conductors and so are used to carry electricity from the outlet through the metal cord (in the insulating cover) to an appliance like a vacuum cleaner or cell phone charger.

The terms *insulator* and *conductor* can also be used to describe how a material conducts heat. Here, we're only talking about electricity.

On a microscopic scale, you can think of conductors as having atoms whose electrons can move within the material.

Microscopically, insulators have *immobile charges*. Their atoms have electrons that don't wander.

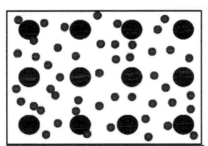

Conductor: The atoms or molecules (red, above) are in a regular arrangement. The electrons (green) move among them. The material doesn't have a charge, even though the electrons can move.

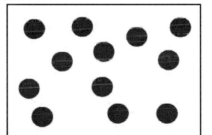

Insulator: The atoms or molecules (orange, above) aren't in a regular arrangement. Electrons aren't shown in this diagram—they stay with their atoms and don't move away.

Most conductors have a regular, or crystalline, array of atoms. This means that their atoms are arranged in a repeating pattern. Any upset to this pattern creates regions in which electrons can't move as well.

Most insulators don't have an orderly arrangement of atoms. They have an amorphous structure (remember chapter 2?), which keeps the electrons from moving within the material.

DOPING ADDS NEW POSSIBILITIES TO SEMICONDUCTORS

The word *gate* is used a lot in this book. There are transistor gates and logic gates. In both cases, though, *something* is controlled. In a transistor, the gate is a physical structure that controls electron flow. A logic gate uses more than one transistor to determine the outcome of a decision—whether electrons flow and to where.

Remember, a semiconductor is an interesting material that acts like a wire at higher voltages and like a rubber pad at lower voltages. So when we talk about *doping* a semiconductor, this isn't the kind of doping that gets people arrested; it's adding controlled amounts of specific materials to pure semiconductors. When you add elements like boron or phosphorus to a whole piece of silicon, it's called *bulk doping* and it lowers the voltage at which the silicon changes from insulator to conductor. When you add these *dopants* in the right amounts to tiny parts of the silicon, you change the voltage gap in that part of it. If you can isolate, or separate, the doped regions from each other, you can create a transistor.

GATES CONTROL ELECTRON FLOW

Each transistor on a silicon chip has a set of conductive regions surrounded by insulating areas. The conductive regions are controlled by a *gate*. The gate of a transistor is different from a logic gate. A transistor gate controls the flow of electrons between these conductive regions by changing the voltage in the semiconductor part of the transistor between them. It acts like the valve in a tube between two containers of liquid, like lemonade. When the transistor is turned on (the valve is open), the gate voltage allows electrons to flow from one conductive, doped region to the other and then out of the transistor into the circuit. When the transistor is turned off (the valve is closed), the gate voltage allows no electron flow and the electrons are contained within the transistor. This ability to switch individual transistors allows the integrated circuit to carry out computing instructions (more about that in chapter 5) or to store data (a 1 or a 0).

INTEGRATED CIRCUITS CONTAIN MANY TRANSISTORS LINKED TOGETHER

In really old computers, transistors were connected with wires. In integrated circuits, they're connected with conductive materials right on the chip. So the integrated circuit behaves like a *system* of pipes and containers whose valves are carefully planned and controlled. Where the electrons flow is critical to how the integrated circuit works, so each transistor's connections are carefully planned, designed, and manufactured into the silicon. How are these regions of conductive silicon and insulation actually created? It takes special processing equipment to create regions that are this small. And to make enough of them to satisfy the needs of cell phone users, it takes a multi-billion-dollar factory.

The containers of lemonade above are connected by a tube with a valve in it. The valve controls the flow of lemonade from one container to the other. In the top two diagrams, the valve is closed, so even if the containers are tilted (middle diagram), the lemonade doesn't flow. In the bottom diagram, the valve is opened, so only then can the lemonade flow.

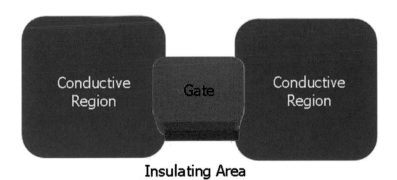

The transistor gate controls the electron flow in the area underneath the insulating layer (between the two conductive regions—orange in the diagram). Controlling this flow controls the transistor's state: whether the transistor is on (1) or off (0). The voltage in the conductive regions is important too. The gate voltage determines whether or not the electrons *can* flow; the voltage in the conductive regions determines *how many* electrons flow.

From Sand to Wafers to Chips

A *clean room* is just that—a room that is very clean. The cleanliness of a room is measured by:

- How big the particles in the air are and,
- How many of them there are

In rooms clean enough for transistor making, the air is filtered so that no particle in it is bigger than half a micron (500 nanometers). Even particles smaller than that can cause great damage to the transistors.

The filters don't get all the particles. The number of particles they leave in the room determines the *class* of the clean room. Most chip manufacturing takes place in a class 10 clean room or cleaner—meaning that on average, only 10 (or fewer) half-micron particles are in a cubic foot of air. By contrast, an operating room is about class 100,000 and a typical office or home is class 1,000,000.

CLEAN ROOMS WITH A VIEW

Sitting atop a mesa, in the city of Rio Rancho, New Mexico, are some of the world's largest clean rooms. They were built to house integrated circuit fabrication plants, or *fabs*. Many of the world's chips are produced at this site. Just one chip produced today has far more computing power than roomfuls of those early computers. How these chips are produced to accomplish such a feat is an amazing story.

MAKING THE TRANSISTORS ON A CHIP IS FAR FROM EASY

To fit the incredible number of transistors needed inside your phone into such a tiny space, chip designers must make each transistor very, very small. In fact, if you cut a single hair on your head

Human hair
17,000 nanometers across

and looked at the cut end, you could fit several *thousand* transistors across it. Humans can't directly touch and handle materials at such small sizes so special procedures and equipment must be used to build transistors—quite literally—layer by layer.

THE CHIP STARTS OUT AS PART OF A WAFER

The entire silicon *wafer* is processed in the fab. Each working chip is separated afterward and put into its own package. To make the transistors' conductive and insulating regions on the chip, the silicon must be changed in one of the following ways:

- Doping different atoms into the silicon
- Reacting the silicon with chemical(s) and heat

These transistor regions then need to be connected to other regions and other transistors by:

- *Depositing* material (typically a metal, which is conductive, or a glass, which is insulating)
- Removing the unwanted material (from areas where it shouldn't be)

In the next few pages, we'll learn about the materials used in making a transistor and how they're put together.

CONTROLLING THE CONDUCTIVE AND INSULATING PROPERTIES OF THE TRANSISTOR REGIONS IS CRUCIAL

That means starting with pure silicon. Pure silicon is refined from sand. Yes, the stuff you walk on at the beach. Sand is mostly *silica*, or *silicon dioxide*. A molecule of silica contains one silicon atom and two oxygen atoms. To get pure silicon from sand, the silica is refined by causing a complex series of chemical reactions between sand and other materials at a high temperatures. The carbon and oxygen atoms

Most transistors today are *25 nanometers* across. The finest human hair measures approximately *17,000 nanometers* across. There are one billion (1,000,000,000) nanometers in a meter; a meter is a little more than a yard.

There are other forms of silica, including quartz, which is crystalline, and opal, which is amorphous. Sand contains other materials than just silica—like finely broken seashells and different kinds of salts. Because sand is so abundant, it's easier to purify the silica from sand than to mine quartz or opal and refine silicon from those sources.

Some of these materials are just high-tech versions of the charcoal you see in the grill. Some are more unusual.

form carbon dioxide molecules, a gas, leaving relatively pure silicon, called *metallurgical silicon*, which is 98 percent pure silicon. But this silicon isn't pure enough for semiconductor transistors—it's refined further to 99.999 percent pure.

How Pure Is Pure?

Metallurgical silicon is at least 98 percent pure, but silicon destined for semiconductor devices is refined to the point of being about 99.999 percent pure. This means that more than 99,999 tons in every 100,000 tons is silicon.

To put this percentage in perspective, there are roughly 300 million people in the United States. If only 3,000 people in the entire United States didn't have green eyes, then you could say that the United States was a 99.999 percent green-eyed country!

THE FIRST IC PROCESSING STEP IS DOPING THE PURE, LIQUID SEMICONDUCTOR-GRADE SILICON

Don't try this at home: silicon melts at a little over 1400 degrees Celsius or 2550 degrees Fahrenheit. Doping, in this case, means adding some *boron* or *phosphorus* atoms to the pure, liquefied silicon. These phosphorus or boron atoms will replace some of the silicon atoms in its regular structure when the silicon solidifies into a single-crystal, cylindrical *ingot*. Since phosphorus atoms have one more electron than silicon atoms, putting them into the silicon crystal results in extra electrons. Similarly, putting boron atoms—which have one fewer electron than silicon atoms—into a silicon crystal results in too few electrons. These extra (or fewer) electrons in the crystal are important for later creating the electron flows in the transistors. After cooling, the ingot is sliced crosswise into circular wafers using enormous but very precise circular saws. The saw blades have a hole in the middle, like a doughnut. The ingot starts in the doughnut hole part, which is where the cutting action happens. After slicing, the wafers are polished, cleaned and shipped to fabs.

THE NEXT STEP IS TO CREATE THE PARTS OF THE TRANSISTORS THAT WILL GENERATE THE ELECTRICAL FLOW

To do this, the silicon must be doped more in certain areas to create even more or fewer electrons there. This is usually done by *implanting ions*, or sticking ions in those certain areas of the wafer. An ion is an atom with a different number of electrons than protons, so it has an electric charge (more on ions in chapter 7). Atoms themselves can't be implanted, but their ions can.

Because they are the result of slicing by saw, some people call wafers "slices."

The entire silicon ingot is one crystal—it has the same repeating pattern throughout. Single-crystal silicon forms a cylindrical shape.

IMPLANTATION IS A VERY PRECISE WAY TO CONTROL THE AMOUNT, TYPE, AND DISTRIBUTION OF IONS EMBEDDED IN THE SILICON

These factors are critical to the transistor because they control the amount of electrons that can flow when it's working. Let's make a model of ion implantation by thinking of stacks of oranges, which represent ordered atoms in the wafer. Instead of the stack of oranges in the store that you can stand next to, imagine a huge room layered in oranges many feet deep. Now imagine suspending yourself on a rope from the ceiling (facedown in a harness like a super-spy in the movies) over the top layer of oranges. Like the stack in the store and like silicon atoms in a wafer, the oranges are in precise positions. There may be an imperfection in the stack here or there (a grapefruit, a plum, or even an empty space—remember, it's only 99.999% pure!), but overall, it's very ordered. In your backpack, you bring your favorite implanting material—limes—and you sneak them into the layers by pushing the oranges aside a little. The limes (which represent boron; phosphorus would be more like an apple in size) can go to different depths depending on how hard you work, and you can count the limes to control the number you insert. You could also control

Notice how precisely the ions are embedded in the silicon in this Play-Doh model. (See the activity at the end of this chapter to learn how to make your own Play-Doh model).

where to insert the limes if you made yourself swing on the rope. And even if you carried a mixed bag of fruit (who packed your backpack anyway?), you could choose just the limes, not the lemons or cantaloupes.

IMPLANTATION LIMITS

Back at life size, we can't choose individual spaces between the silicon atoms in the wafer for implanting and we can't move the ions into the wafer one at a time. The ions are actually "shot" at the wafer in a stream. Through some clever (and complex) engineering, the type of ions shot, how deep they go, and how many are implanted can be precisely controlled. To prevent strays from going where they shouldn't (and wreaking havoc there), a layer of material is used to protect the areas that shouldn't get the implant. This material acts like a stencil, preventing the ions being implanted from reaching the silicon surface. Unlike the lettering stencils in an art store, this stencil must be many, many, many times smaller.

THE STENCIL MATERIAL USED IN SEMICONDUCTOR PROCESSING IS CALLED PHOTORESIST

Why? Because light is used to harden it. *Photoresist* is a liquid that is spun onto the wafer to coat it evenly. Then some areas are hardened by shining light onto them while others aren't. A *mask* is used to control where the light reaches—it creates shadows on the wafer. The areas that aren't hardened can be washed away easily. In this way, a stencil is created right on each wafer every time it's needed. Each time after the hardened photoresist has done its job, it must be completely removed. Sometimes, chemicals are used to remove hardened photoresist; other times, a combination of chemicals and *plasma* is used. Photoresist does not become part of the chip.

The light used for photoresist hardening has a very short *wavelength*. Our eyes cannot see this light. More on wavelengths in chapter 6.

ONCE THE CONDUCTIVE AREAS OF SILICON HAVE BEEN MODIFIED, THE GATE IS CONSTRUCTED

The gate controls the electron flow between the conductive areas. Remember, controlling this electron flow is critical—it's what determines whether the transistor stays on or off when it's switched.

The gate itself is typically *polycrystalline* silicon (often called *polysilicon*) and is conductive. It's separated from the wafer by a

Plasma, in this case, refers to a high-energy gas-like substance that has lots of free-moving ions and electrons. It's very reactive.

The transistor gate (green) and gate oxide (blue) in a Play-Doh model above the region between two conductive areas (purple).

Polysilicon has regions of regularly arranged atoms all going in different directions. Again, *poly* means "many," so polycrystalline silicon has many crystals of silicon.

This thin layer of insulating glass—the gate oxide—is made by reacting oxygen with the silicon on the surface of the wafer. This reaction creates silicon dioxide, or silica, just like the sand the wafer came from!

thin layer of insulating glass called the *gate oxide*. Making this thin layer of glass takes precise control of the conditions in a special furnace (similar to an extremely hot oven). The wafer is put into this furnace with oxygen gas, which reacts with the silicon, turning the outer surface of the wafer (the top several layers of oranges in our huge room) back into silicon dioxide. Like beach sand, this silicon dioxide doesn't have ordered molecules; it's amorphous. How fast the thickness of the gate oxide grows depends on how hot the oven is and how much oxygen is available.

To make the gate, silicon atoms are deposited over the entire wafer and then removed from where they shouldn't be—everywhere that isn't the gate. Removing the polysilicon is done by *etching* it away—eating away the polysilicon by reacting it with *acid* or plasma. As before, photoresist is used as a stencil to control where reactions occur and where material is deposited.

ONCE THE GATES ARE COMPLETED, THEY NEED TO BE CONNECTED TO THE OTHER PARTS OF THE TRANSISTOR

The transistors are then connected to each other. Since we can't attach wires where we need them, these connections need to be made by:

- Depositing layers of insulation

- Etching the insulation away—cutting through those insulation layers in certain places

- Depositing layers of metal (aluminum, copper, or something more exotic)

- Etching away the metal where it isn't needed to form the connections or wires between the transistors and other connections

These steps are repeated to build multiple layers of connections, and each time, the photoresist is used as a stencil to guide the process of deposition and etching.

INSPECTIONS TEST THE TRANSISTORS AND THEIR CONNECTIONS

Once the wafer processing has been completed, each chip on the wafer is tested to make sure that the transistors and their connections work properly. After this inspection, the wafer is *scored*—precisely scratched—to separate the chips from each other. Each working chip is placed inside of and attached to a *package* and then is tested again. Most packages are made out of plastic and look like bugs with metal legs called *leads*, but packages do come in many shapes and sizes. The leads are connected to precise areas on the edges of the chip with thin gold wire and carry electrons to and from the chip inside. The packages can then be attached to a plastic *circuit board*

Take another look at the picture of the phone circuit board at the front of this chapter. The black rectangles and squares are packages with chips in them. Do any of them look like bugs to you?

and connected to each other to build a computer, a cell phone, an electronic game, or almost anything else that runs on electronics.

TRANSISTORS HAVE DIFFERENT USES ON THE CHIP

Some transistors on the chip store information, so they aren't changing all the time. Some chips, in fact, are designed specifically for storage, or memory. Others are used in the computing that a chip does and must change often and quickly when you use your cell phone. The faster these transistors can change their state—that is, change from a 1 to a 0 or vice versa—the more instructions the chip can carry out per second and so the more computing power the chip has. Now you know that in order to change how a transistor functions, you need to change the gate, conductive regions, or connections.

Today we have chips with over one billion (that's 1,000,000,000) transistors. The computer in your cell phone has more computing power than roomfuls of minicomputers! But all these transistors don't work by themselves—they require well-written software.

Achieving faster transistors depends on making them in smaller sizes and making better connections between them. The semiconductor-processing industry has done this over the last 45 years or so and continues to do so today. In fact, the question in the industry is, "When will transistors stop shrinking?" In 1965, Gordon Moore predicted that the number of components on an integrated circuit would double about every two years. This prediction has largely come true.

By the way, even Gordon Moore has at times expressed surprise at how well his prediction has held up!

Build a Transistor out of Play-Doh

To really get a feel for the limitations of working with such small sizes, try this activity to build a transistor!

You'll need the following materials and equipment for this activity:

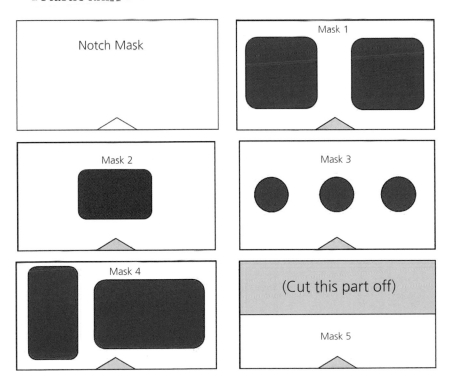

- ❂ Play-Doh in five different colors: white, yellow, green, blue, red
- ❂ Very small dark-colored beads
- ❂ Paper copy of the masks below, doubled in size and with the colored sections cut out
- ❂ Plastic knife

Notch Mask	Mask 1
Mask 2	Mask 3
Mask 4	(Cut this part off) / Mask 5

1. MAKE THE SILICON WAFER

Instead of photoresist masks, we use paper masks in this model.

❧ Press out the white Play-Doh to a ½-inch-thick oval.

❧ Cut a straight edge into it.

❧ Line up the notch mask along the straight edge and cut the notch into the edge.

2. IMPLANT THE CONDUCTIVE REGIONS

❧ Place mask 1 on the wafer, being careful to align the point of the notches on the mask and in the wafer.

❧ Spread the beads (boron or phosphorus) over the mask and white Play-Doh.

❧ Push the beads into the white Play-Doh with your finger tips (conductive regions).

❧ Remove mask 1.

Notice that the beads are pressed into the wafer only in the regions not covered by the mask. On an actual wafer, you wouldn't be able to see where the implanted ions are.

3. GROW THE GATE OXIDE

In actual production, layers of oxide or metal are either grown or deposited onto the wafer. So, here is one place where this model of semiconductor manufacturing breaks down— a separate layer of material that has to be aligned onto the wafer isn't created at all in actual production.

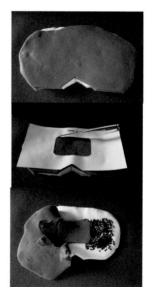

❧ Form the blue Play-Doh into a ball and press it out to ⅛-inch thick. Cut off one edge and put a notch into that edge (like you did with the white Play-Doh before).

❧ Place the blue Play-Doh on the wafer, lining up the notches.

❧ Place mask 2 on the wafer, lining up all the notches exactly.

❧ Cut around the edge of the shape in mask 2, being careful to cut through only the blue layer.

❧ Remove mask 2.

❧ Remove the excess blue layer, leaving only the rectangular shape (gate oxide).

ACTIVITY

4. DEPOSIT THE METAL GATE

- Form the green Play-Doh into a ball and press it out to ½ inch thick. Cut off one edge and use the notch mask to put a notch into that edge as before.

- Place the green clay on the wafer, aligning the notches.

- Place mask 2 on the wafer again, lining up all the notches exactly.

- Cut around the edge of the shape in mask 2, being careful to cut through only the green layer.

- Remove mask 2.

- Remove the excess green layer, leaving only the rectangular shape, which should be sitting directly on top of the gate oxide (blue). This is the metal transistor gate.

5. DEPOSIT AN INSULATING LAYER

- Form the red Play-Doh into a ball and press it out to 1 inch thick. Cut off one edge and use the notch mask to put a small notch into that edge.

- Place the red Play-Doh (insulating layer) on the wafer, lining up the notches and using your fingers to press the clay around the gate and gate oxide.

Notice that the surface of the wafer is no longer flat. This unevenness affects how the next layer is deposited, and it happens on a real wafer too.

The effect is similar to a light snow falling on uneven ground—you see some rocks sticking out from the snow. On a wafer, letting the underlying layers peek out from the deposited layer affects how the chip works. So, the design of the chip must balance between depositing a lot of material to get good coverage and trying to keep the surface as flat as possible.

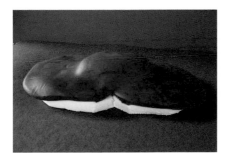

The insulating layer is usually silicon dioxide, silica, that is deposited onto the wafer.

ACTIVITY

6. MAKE THE CONTACT HOLES

The contacts connect the metal layer, which acts as a wire, to the conductive regions and the transistor gate. On a real wafer, this metal would be deposited and the excess removed. Getting the metal to deposit all the way into the contact holes can be difficult.

- Place mask 3 on the wafer, lining up all the notches exactly.
- Remove the insulating layer (red Play-Doh) from the circles in mask 3. These are the contact holes.
- Remove mask 3.

7. MAKE THE CONTACTS

- Separate a small amount of yellow Play-Doh and press this into three small balls each the same size as the contact holes.
- Insert the yellow Play-Doh balls (metal contacts) into the contact holes. Fill the holes as completely as you can without disturbing the insulating layer.

8. DEPOSIT THE METAL LAYER

- Form the remaining yellow Play-Doh into a ball and press it out to ½-inch-thick. Cut off one edge and use the notch mask to put a small notch into that edge.
- Place the yellow Play-Doh (metal layer) on the wafer, lining up the notches and using your fingers to press the clay onto the uneven surface of the wafer.
- Place mask 4 on the wafer, lining up the notches carefully.

- Cut around the rectangular shapes, being careful to cut through only the yellow Play-Doh (metal layer).
- Remove mask 4.
- Remove the excess metal layer, leaving the two rectangular shapes.

ACTIVITY

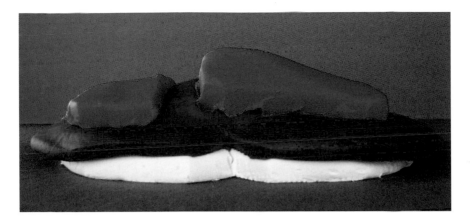

CONGRATULATIONS! YOU'VE COMPLETED YOUR TRANSISTOR. TAKE A LOOK INSIDE:

- Place mask 5 on the wafer, aligning the notches carefully.
- Cut through your device as marked on mask 5.
- Compare the cross section of your device with this photo. It should look similar!

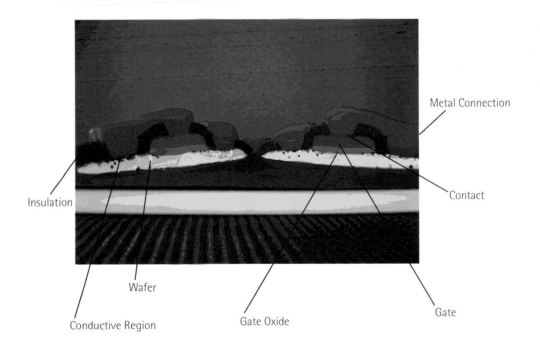

Metal Connection

Contact

Insulation

Wafer

Conductive Region

Gate Oxide

Gate

CHAPTER 5

A Cell Phone's Logic Is Written in Its Software

Software is what gives your cell phone the ability to do the amazing number of tasks that it does. An e-mail, a text message, a photograph, an Internet page, and a phone call—they're all 1s and 0s when they're flying through the air. It's the software that knows the difference.

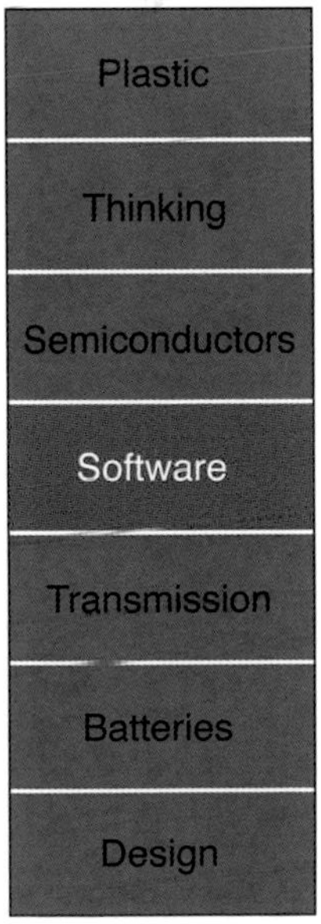

PROGRAMMING WELL MEANS USING YOUR RESOURCES WELL

The topic of software programming is huge, and workers in the field have many opinions on how software should be designed, written, tested, and put to use. We can't cover everything about software here, but we will talk about why writing good software is important and how your cell phone uses well-written software to do all the things that it does. It comes down to using resources well.

Plastic

Thinking

Semiconductors

Software

Transmission

Batteries

Design

MEMORY IS A RESOURCE

Today's chips contain many millions of transistors. As you learned in chapter 4, each transistor can store a 1 or a 0 and now you know *how* the transistor does this. But whether you have just 10 or as many as 10 million transistors, you run out of storage space fast if you don't use them efficiently. Storage places, or memory, is one of the resources a computer uses.

```
00001110100100101010001111
00111110010010100110101
11010011101010011110010
01100001101010010100010
11010000101111101011101
00101101110000000101010
```

TIME IS A RESOURCE

Just storing the 1s and 0s doesn't do much; you also need *instructions* on what to do with them. From chapter 3 you know that simple logic gates can be combined to express very complex decisions like what vehicle to use in certain weather conditions. Each logic gate is simply a set of transistors connected in a specific way. So how fast the computer *executes*, or carries out, instructions depends on how fast the transistors can switch from a 1 to a 0 and vice versa. And how fast the computer can execute the instructions you give it can mean the difference between getting your e-mail today or next week (or even later!). But here, too, being efficient in writing instructions is key.

The instructions you give your computer have to turn into instructions your computer understands. Knowing how the software works with the hardware is important to managing the time it takes to execute your instructions.

Memory can be used to store data or instructions. How much memory a chip has for each depends on how it's designed. We'll look at this first . . .

Memory

WHAT IS MEMORY?

We said in chapter 4 that transistors store 1s and 0s and that these 1s and 0s can then be used to:

- Build logic gates, which can be combined to build complex instructions, or;
- Store data that the instructions act on.

To do these things, two main kinds of memory are needed: *volatile* and *persistent*.

YOUR CELL PHONE USES DIFFERENT TYPES OF MEMORY

Your cell phone uses *flash* and ROM memory chips for its persistent memory. Flash memory can be *overwritten*; in other words, the transistors in them can be changed if you want to store different data. ROM memory can't be changed because the chip is programmed during its manufacture and stays that way.

Flash memory is used to store your information. Think about saving the last 10 numbers that called your cell phone. Once you get the 11th call, the first of the 10 numbers is deleted—the memory is overwritten.

ROM memory is used to store things like how to start up your phone and how to search for a signal, which are things your phone will always need to be able to do. Nowadays, flash memory is used for a lot of these kinds of functions too—that way, the software in your phone can be updated without having to replace the chips. For example, if your cell phone service provider changes its

Volatile memory requires a steady flow of electricity to maintain. Once the electron flow is cut off (you unplug your computer or your phone battery dies), the data are lost—that is, the transistors switch off.

Persistent, or nonvolatile, memory doesn't require a steady electron flow. Once the transistors have been switched to 1 or 0, they stay that way.

When you use your cell phone to create a text message, for example, you're using *RAM* (random-access memory). This is volatile memory—the transistors change—and you know this because the software you use changes the display on your screen (that is, it shows the letters you type). But if your battery dies while you're writing your text message, your data— the text message you were working on—are gone.

If you want to save that text message, you store it to the flash memory chip in your phone. A flash memory chip is persistent memory—it stores your data even if the power to it is turned off.

Your cell phone has *ROM* (read-only memory), on which is stored the instructions to *boot*, or start, it up. This memory is always there and doesn't change even if the phone is turned off.

network (more about that in chapter 6), it can update your phone with these changes quickly and easily so you don't lose service. Think of your phone as having two banks of flash memory—one for your phone's functions and one for your information.

Software Must Work with Hardware

SOFTWARE TO HARDWARE TO SOFTWARE AGAIN

We said before that knowing how software and hardware work together is important to managing the time it takes to execute your instructions. What does this mean? Well, imagine trying to talk with

someone who didn't speak your language. Any talking would have to be done through a translator. Having your phone's computer execute your instructions is similar—you and your phone's computer don't speak the same language.

So, to get your phone to do what you want (send an e-mail, answer a call), you must find a way to give it instructions it can understand. Then your phone's computer has to have a way to give *you* an answer you can understand. In other words, your instructions must be translated for the computer so it can process your data, and then the answers it gets must be translated back into your language.

LAYERS OF ABSTRACTION SIMPLIFY OUR WORK

We mentioned in chapter 3 that abstraction helps us to interpret the world around us. Let's think about this idea in a different way now. For example, think about the stages you went through to learn how to read:

- First, you (probably) learned to recognize each letter of the alphabet.
- Then you learned what sound each letter makes.
- Then you learned how to combine letters to make words that represent something.
- Then you learned how words combine to make sentences that convey meaning.
- Finally, you learned that sentences can be combined into paragraphs, which can be strung together into essays, books, stories, and so on.

Each of these stages is a *layer of abstraction*. Each layer lets you to make a leap in learning: if you don't have to think about what sound the letter *a* makes, then you can just concentrate on the meaning of what you're reading.

"My dog ate my homework!"

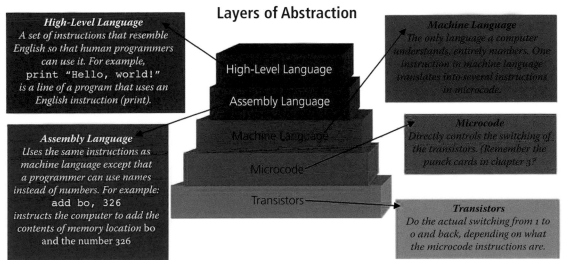

Layers of Abstraction

High-Level Language
A set of instructions that resemble English so that human programmers can use it. For example,
`print "Hello, world!"`
is a line of a program that uses an English instruction (print).

Assembly Language
Uses the same instructions as machine language except that a programmer can use names instead of numbers. For example:
`add b0, 326`
instructs the computer to add the contents of memory location b0 and the number 326

Machine Language
The only language a computer understands, entirely numbers. One instruction in machine language translates into several instructions in microcode.

Microcode
Directly controls the switching of the transistors. (Remember the punch cards in chapter 3?)

Transistors
Do the actual switching from 1 to 0 and back, depending on what the microcode instructions are.

High-Level Language
Assembly Language
Machine Language
Microcode
Transistors

Remember we said at the end of chapter 3 that one of the benefits of separating hardware from software is that software can be made to run on any computer—making the effort to create it worthwhile?

Another benefit—for the programmer—is that once you learn a high-level language, like C, you can write programs for any computer that can translate that language. So, a programmer doesn't have to learn a different language for each computer—learning just a few high-level languages pays off.

Of Programmers and Code . . .

- A program, or code, is a set of instructions.

- A programmer is someone who creates programs.

- The programs a programmer writes are usually referred to as source code.

Remember in chapter 3 we talked about logic gates? Well, they're here—logic gates are the instructions we're referring to at the microcode and transistor levels.

A COMPUTER USES LAYERS OF ABSTRACTION TOO

The model above uses five layers of abstraction to describe the different stages of translating your instructions into instructions your cell phone's computer can understand. Each layer lets you make a leap in translation. We'll learn about each layer starting with the top—the human end.

A Programmer Works at the Highest Level of Abstraction

By programming in a *high-level software language*, the programmer thinks about how to create a sequence of instructions that will be executed as quickly as possible. Programmers don't think about how each transistor is switched—and they don't need to. Those details are handled by the other, lower layers of abstraction, which we'll learn about below.

A PROGRAM CAN BE RUN IN DIFFERENT WAYS

How is a program—source code—converted into language that a computer understands? The answer depends partly on which high-level language the source code is written in. Most of the code for cell phones is written in a language called *C*, which requires a *compiler*. Other languages, like *BASIC*, use an *interpreter*. Some programs are written in *assembly language* and require an *assembler*.

A compiler is a program that collects and rearranges all the instructions in the source code to make it execute faster. This means that the compiler must execute first: it collects all the instructions in the source code, rearranges them, and creates an object code—a set of instructions translated into machine language—which the computer can then execute.

An interpreter converts each instruction in the source code and executes it immediately. It doesn't collect and rearrange all the instructions. So, the program usually runs more slowly because each instruction is translated through all the layers of abstraction one at a time.

An assembler converts source code written in assembly language to machine language. Both use the same instructions, but programmers can use names and letters in assembly language, making programming easier.

Most chips use only one machine language but can translate an almost infinite number of high level languages.

EACH CHIP HAS ITS OWN MICROCODE

Once the source code is translated into machine language, the computer's chips convert the machine language instructions into microcode. Each kind of chip has its own microcode, which is *hard-wired*—it's physically designed into the chip's transistors and can't be changed. Because of this hardwiring, each chip requires its own version of machine language, assembly language, and compilers. Still, software can be executed on different computers as long as the chip has a compiler, interpreter, or assembler for that language.

MICROCODE, TRANSISTORS, AND INSTRUCTIONS

At the lowest level, the original source code is now microcode switching transistors from 1 to 0 or vice versa. But how does this transistor switching work? And, how does the cell phone's computer keep track of what all those transistors are doing? We'll need to get back into the hardware (briefly) to answer those questions, and we'll start where we left off in chapter 4.

MEMORY MANAGEMENT: HOW THE CHIP KEEPS TRACK OF DATA

In chapter 4, we described how a transistor is made and how wire-like connections can be created on the chip. These connections act as pipes, only they carry electrons instead of water.

Earlier in this chapter, we said that your cell phone's computer uses memory to store data and instructions. Having a lot of storage is good, but your cell phone's computer must be able to use that data and those instructions when it needs to, and that means transferring 1s and 0s from ROM or flash into the cell phone's RAM.

YOU CAN THINK OF RAM AS AN ARRAY . . .

of valves and containers connected to a grid of pipes (Remember our lemonade model of transistor gates in chapter 4?).

IN REALITY . . .

The pipes are metal connections that carry electrons. At each intersection of the metal grid is a transistor, which acts as a valve. Connected to each transistor is a *capacitor*, which holds electrons (like a container holds water or lemonade).

Quick summary . . .

In chapter 3, we saw that logic gates are the basic building blocks of decisions and can use inputs as simple as two transistors.

In chapter 4, we described how transistors are made, connected, and controlled.

In chapter 5, we see that at the microcode layer of abstraction, the original source code is a set of instructions that control which transistors switch on or off.

Data Lines

Address lines are the RAM's way of tracking where a number is.

Data lines express that number in *binary*—using just 1s and 0s.

Each 1 or 0 is called a *bit*.

Address Lines

A set of 8 bits is called a *byte* and translates to a decimal number up to 255.

For example, 57 is:

111001 *expressed in binary*

00111001 *expressed as a byte*.

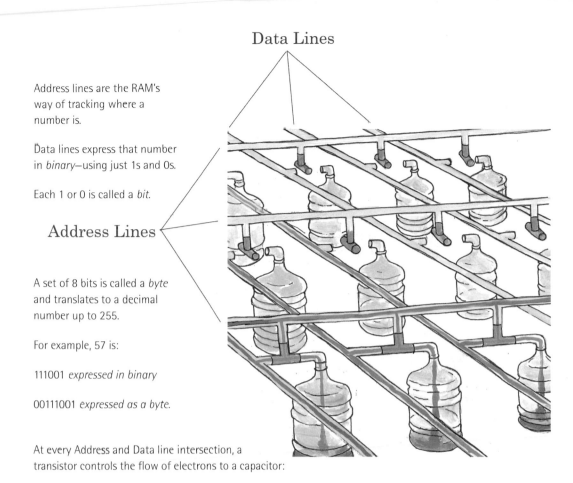

At every Address and Data line intersection, a transistor controls the flow of electrons to a capacitor:

If the transistor (red) is OFF: the capacitor (yellow) cannot charge or discharge.

If the transistor is ON: the capacitor (yellow) charges or discharges through the transistor (red)

Since moving data and instructions from ROM or flash to RAM involves retrieving and storing the data—the 1s and 0s—let's look at how the data gets in and out of RAM.

TO STORE DATA

To store data, electrons flow into the address line (A1, blue) and into the data lines that correspond to 1s (D3 and D7, also blue). This switches the transistor at that intersection on, which then charges the capacitor. The capacitor holds the charge until it is discharged.

So the data written to A1 is: 01000100

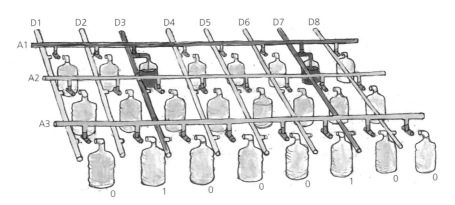

TO RETRIEVE DATA

Electrons flow into the address line from which the data are to be read (A2 this time, red). The capacitors that are charged along this address line (only one at D5 in this case) discharge—the electrons stored in them move out. The voltages along these lines then switch other transistors, depending on how the microcode is written.

So the data read from A2 is: 00010000

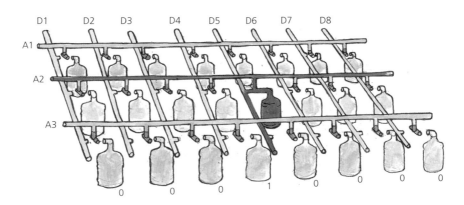

WHEN THE DATA ARE RETRIEVED, YOU CAN THINK OF THEM AS BEING PLACED ON A PILE

Once in RAM, data are manipulated according to the microcode instructions (which are translations from the original source code, remember?). Data are read and processed, and the results are stored to be used in other processing or to be displayed. There aren't any physical piles in the RAM, just areas where transistors switch on or off. But the image of a pile is a useful model to help us understand how the microcode manipulates the data.

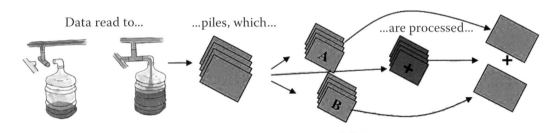

THERE AND BACK AGAIN

Now you know how instructions in software are translated into the 1s and 0s your cell phone's computer understands. The RAM in your cell phone's computer gets the instructions (or software translated into microcode) and data (more on where that comes from in chapter 6) and processes the data using those instructions. Now, to give you the results of its processing, like a text message, the microcode translation also tells your cell phone's computer to display the results in a particular way—a way that you will understand, like a bunch of words on a screen. But let's get back to why well-written software is important.

Another way to say *store data* is: *write data to memory.*

Another way to say *retrieve data* is: *read data from memory.*

SO, WHAT DOES "PROGRAMMING WELL" MEAN?

We said in the beginning of this chapter that programming well means using your resources well. Now that you know how memory is stored and read and how instructions are translated, here's an example to illustrate the importance of good programming:

We have one million numbers, in a completely random order, that need to be sorted from the lowest number to the highest.

Question
Which is faster, a PC or a supercomputer?

Answer
It depends!

above: Personal Computer (PC): Executes 1 MIPS (million instructions per second)

right: Supercomputer: Executes 100 MIPS (million instructions per second)

ALGORITHMS ARE WAYS TO DO THINGS

An *algorithm* is simply a method or rule by which to do something. The key to software programming is to identify the problem to solve, break it down into its logical parts, and then choose the right algorithms to solve each part. So in order to solve the sorting problem above, we need to think about how information is stored and instructions are executed. For example, you could have a rule that said: *Start with the first number and insert the second number before or after it. Then insert the third number in place and continue until all the numbers are sorted.* This rule is similar to how you might sort the cards in your hand while playing a card game. Another algorithm would be: *Divide the entire group into ordered pairs, then combine the pairs into ordered groups. Keep combining the ordered groups until all the numbers are sorted.* Actually, these two algorithms have names—insertion sort and merge sort—and are shown below with a smaller set of random numbers (10 numbers instead of 1,000,000).

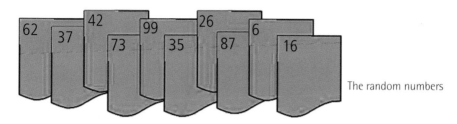

The random numbers

Insertion sort

ALGORITHM 1: INSERTION SORT

Start with the first number and insert the second number before or after it. Then insert the third number in place and continue until all the numbers are sorted.

Step 1: Start with the first number (red) and insert the second number (yellow) before or after it.

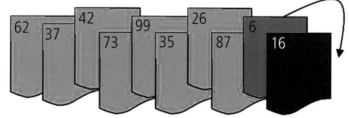

Step 2: Insert the third number (blue) in place.

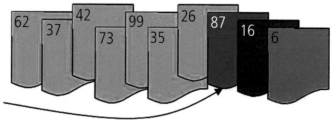

The third number stays where it is.

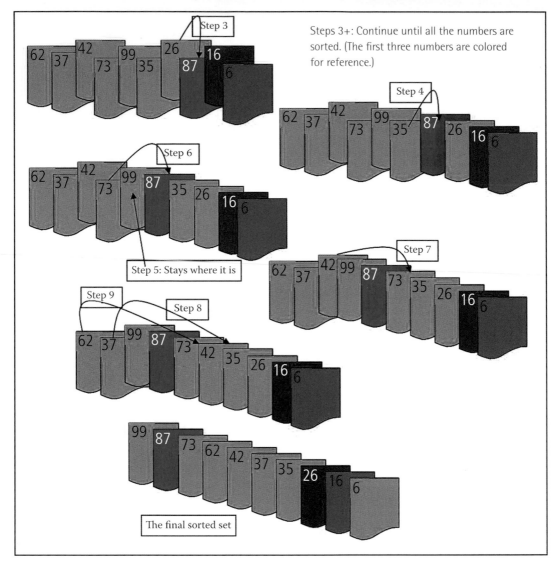

Step 3

Steps 3+: Continue until all the numbers are sorted. (The first three numbers are colored for reference.)

Step 4

Step 6

Step 5: Stays where it is

Step 7

Step 9

Step 8

The final sorted set

Merge Sort

ALGORITHM 2: MERGE SORT

Divide the entire group into ordered pairs, then combine the pairs into ordered groups. Keep combining the ordered groups until all the numbers are sorted.

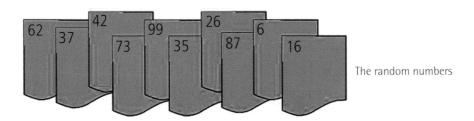

The random numbers

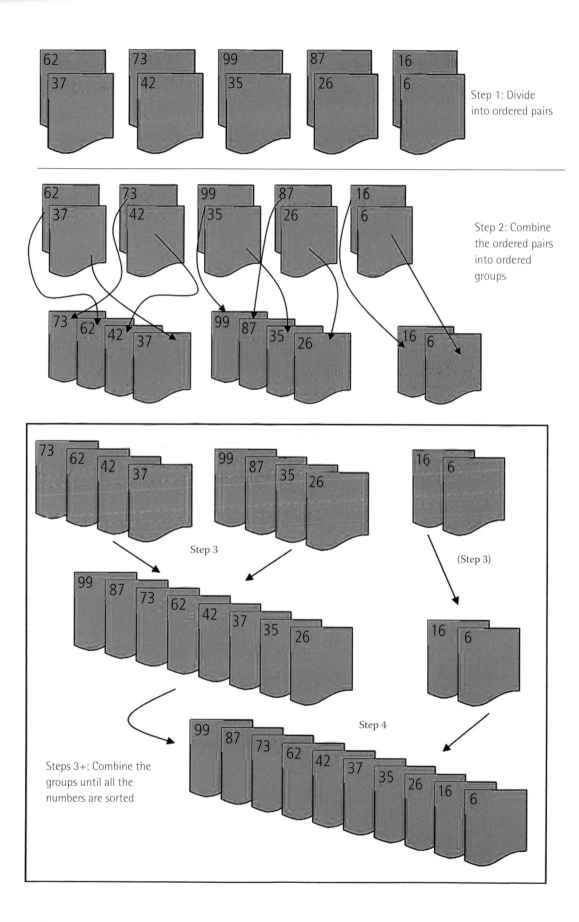

Step 1: Divide into ordered pairs

Step 2: Combine the ordered pairs into ordered groups

Step 3

(Step 3)

Step 4

Steps 3+: Combine the groups until all the numbers are sorted

In this example, we wanted to see how quickly we could sort a large number of items. Merge sort is definitely faster than insertion sort, as we've seen, but insertion sort uses less memory to process the task. (This is not as easily seen in the diagrams above but is true.)

If we wanted to sort a large number of items using a small amount of memory—and it didn't matter as much how long our sorting took—insertion sort might be the better algorithm to use.

THE NUMBER OF STEPS CAN REALLY INCREASE DEPENDING ON WHICH ALGORITHM YOU CHOOSE

Notice the difference in the number of steps each of these algorithms needed. The insertion sort method required nine steps. The merge sort method required four steps, or *fewer than half* the number of steps. Now, we used only 10 items to show these two methods—what if we had to sort more items? How many more steps would be required to sort many, many more items (like one million items)? Instead of trying to actually sort that many items, it's easier to look at the illustrations of both algorithms and draw conclusions. We can see that we needed to move each of the items into the right place in the insertion sort method. In the merge sort method, we could have sorted an additional six items in the same number of steps! The merge sort algorithm is far more efficient *at this kind of task.*

SO WHAT'S THE ANSWER: PC OR SUPERCOMPUTER?

As we said before, the answer depends. The answer depends on which computer used which algorithm. To sort 1,000,000 items, the number of steps required by the insertion sort method is about 1,000,000. The number of steps required by the merge sort method is about 20. That's a *huge* difference. If the supercomputer—remember, it's 100 times faster than the PC—used the insertion sort algorithm and the PC used the merge sort algorithm, the PC would still win, and by a very wide margin, too.

YOUR CELL PHONE'S COMPUTER TRACKS AND COUNTS CONTINUALLY

It tracks the numbers in your phone book; it keeps your e-mails; it catalogs your pictures. It does all this at the 1s-and-0s level, so it handles millions of instructions. So how well your cell phone is programmed is key to how well it works for you.

CHAPTER 6

Getting the Message Through: Call Transmission

The great thing about a cell phone is you can use it to communicate from almost anywhere to almost anywhere. The cell phone can communicate without being wired or otherwise connected to anything. This is truly an amazing, almost magical feat. This chapter describes how wireless communication takes place.

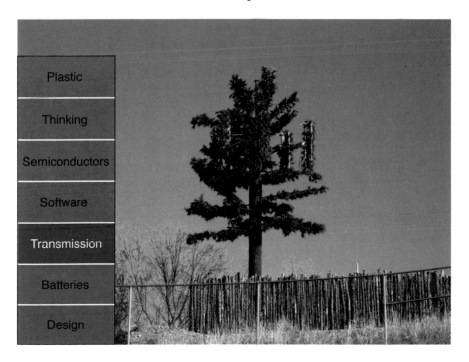

Plastic

Thinking

Semiconductors

Software

Transmission

Batteries

Design

color. On the fourth night, then, even though there are other red-colored tent lights, your friend can still pick out your message because he knows roughly where your tent is and which color and code you're using.

HOW CAN YOU KNOW WHEN TO TALK BACK?

So far in this campsite example, the messages have all been *one way*. That is, the messages were sent without the signaler intending to get a return message. Radio stations operate this way—you hear what they broadcast but don't send a signal back to them. So how could you and your friend have an after-dark conversation from your tents? You would need to use a set of rules to tell each other when your end of the conversation was finished. When you received the signal that your friend had finished using his flashlight, you could use your flashlight knowing that he was watching for your signals.

PROTOCOLS CAN LET YOU KNOW WHEN TO SEND AND WHEN TO RECEIVE

The set of rules that lets each flashlight-using camper know when it's time to send, when it's time to receive, and what code to use to send the signal is called a *protocol*. Protocols are the rules of etiquette of the conversation. Without them, the conversation can't take place. Protocols establish all the rules of any exchange, including what steps to take in order to establish a connection, how to start and end the conversation, who receives and sends (and when), even what kind of information is being transmitted. Every long-distance form of *two-way communication* needs protocols, and every form of long-distance communication has its own set of protocols. Telegraph, telephone, and even smoke signaling, rely on protocols to let

We won't cover protocols in detail here. In the United States, the Federal Communications Commission (FCC) allocates the frequencies for different users. Each industry has its own standard protocols for *all* forms of wireless communication. Many websites and books explain the details of all the different protocols. Other countries have their own versions of the FCC.

the caller and receiver know when to send a message, when to receive a message, how to code and decode the message, and when the conversation has ended. The protocols for telephone in particular are so familiar to us that we don't often realize that busy signals, dial tones, and off-the-hook tones are part of the protocols.

The Magic of the Wave

CELL PHONES SEND AND RECEIVE RADIO SIGNALS

Radio signals are *radio waves* that are changed to create patterns. Your cell phone's computer, which has the cellular communications protocols written into it, can interpret these patterns. This means that cell phones do something similar to the campers: they send a pattern in radio waves to a receiver and receive a pattern in radio waves that they interpret as information. The details of the protocols are very complex, but we'll talk about a few important concepts so that you understand more about how cell phones communicate. First, let's look at how radio signals are produced and sent. Then we can see how the different ways of changing the waves fit different protocols.

A RADIO WAVE IS BOTH ELECTRICAL AND MAGNETIC

When electricity flows through a conductor, like an *antenna*, it creates both an *electric* and a *magnetic field* around that conductor. Electric and magnetic fields are both *force* fields—that is, they can exert force on an object within the space of that field. When one of these fields changes—grows bigger, for example—the other field changes too. It's easiest to change the electric field by changing the flow and direction of the electricity in the conductor, the antenna. Changing the electric field changes the magnetic field. The two changing fields create a disturbance, which radiates away at the *speed of light.* These

If the force is sufficient to move that object, the object will move along the field lines.

Electro-magnetic wave speed is represented by the letter *c* in physics and in many other scientific fields because the constant speed of light is important in many physics theories.

disturbances are somewhat like the surface waves that are created by each drip from a leaky faucet into a sink full of water. We call these disturbances *electromagnetic waves*. Radio waves are just one kind of electromagnetic wave.

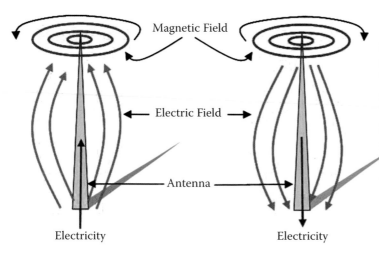

Magnetic Field

Electric Field

Antenna

Electricity

Electricity

Field lines show the direction of the force. They never cross over each other.

You can't see electric or magnetic fields, but they exist, and you can observe the effect they have on objects, like iron filings.

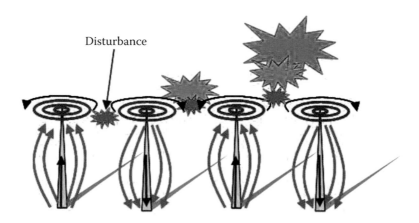

Disturbance

When electricity flows in one direction into the antenna, it creates both an electric and a magnetic field around it. The direction of the electric field is always *perpendicular* (at a 90-degree angle) to the magnetic field. When the electricity flow reverses direction, the electric and magnetic fields reverse direction too.

When the fields reverse direction, a disturbance is created. This disturbance radiates away into space. The pattern of these disturbances is the waves we (or our equipment) detect.

ELECTROMAGNETIC WAVES DIFFER FROM WATER SURFACE WAVES

Electromagnetic waves:

- Don't need to travel through anything. They can even travel through the near *vacuum*, or almost complete emptiness, of outer space. Electromagnetic waves are the result of changing force fields, not the movement of molecules;

- Travel at a constant speed (distance per time): 186,000 miles, or just under 300,000 kilometers, per *second*. This is the speed of light

EVERY WAVE HAS A WAVELENGTH AND AN AMPLITUDE

Wavelength is the distance between one peak of the wave and the next. Smaller wavelengths mean that more waves pass a certain point as the wave travels—in other words, the wave has a higher *frequency*. *Amplitude* is the distance between the highest (peak) and lowest (valley) points of the wave. So, changing the frequency or amplitude of an electromagnetic wave means that the electric and magnetic fields change too. And this means that the electric current through the antenna must be changed to change the frequency or the amplitude of the electromagnetic waves.

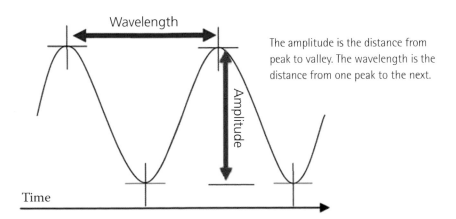

The amplitude is the distance from peak to valley. The wavelength is the distance from one peak to the next.

Humans can see some frequencies of electromagnetic waves. Surprise! we call this range "visible light." We can't see radio waves—their frequencies are too low—but we can still use them to communicate.

We can see frequencies from 4.3×10^{14} to 7.5×10^{14} hertz. Hertz, abbreviated Hz, is a measure of frequency. It's waves per second. So, 4.3×10^{14} Hz means that 430 trillion (430,000,000,000,000) waves move past a point in one second.

Wavelength is related to frequency. Since we just found out that all radiation travels at a constant speed, c, we can calculate any wavelength if we know the frequency of the wave and vice versa. If we represent wavelength as the lowercase Greek letter omega, ω, and frequency as the lowercase Greek letter lambda, λ, then

$$c = \omega\lambda$$

In words, the equation above means that the speed of light is equal to the wavelength of the wave multiplied by its frequency.

$$\lambda = c/\omega \text{ and } \omega = c/\lambda$$

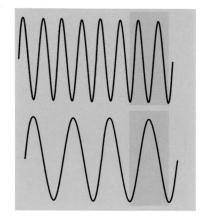

The frequency of a wave is the number of peaks that pass a certain point in a certain amount of time. The grey bar at right shows the same amount of time. The smaller the wavelength, the higher the frequency.

Wavelength

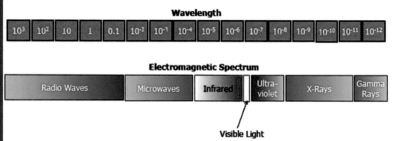

Electromagnetic Spectrum

Radio Waves | Microwaves | Infrared | Ultra-violet | X-Rays | Gamma Rays

Visible Light

In the diagram above, the wavelength decreases from left to right, as shown by the scale on top. Notice that the range of visible light is very small compared with the full range of electromagnetic waves. Since wavelength decreases as frequency increases, the electromagnetic radiation at the right of the diagram (gamma rays, x-rays) has higher frequency than radio waves, which have the lowest. Higher-frequency waves also have higher energy, so microwaves have more energy than radio waves.

Band Designation	Frequency Range	Frequency Use
Extremely High Frequency (EHF)	30 GHz to 300 GHz	Radar
Super-High Frequency (SHF)	3 GHz to 30 GHz	
Ultra-High Frequency (UHF)	300 MHz to 3 GHz	TV Broadcast
Very High Frequency (VHF)	30 MHz to 300 MHz	FM Broadcast
High Frequency (HF)	3 MHz to 30 MHz	Amateur Radio
Medium Frequency (MF)	300 kHz to 3 MHz	AM Broadcast
Low Frequency (LF)	30 kHz to 300 kHz	Radio Navigation
Very Low Frequency (VLF)	30 kHz and lower	Human Hearing

Here are some examples of how the electromagnetic spectrum is used to transmit information in the United States.

GHz = gigahertz, or 1 billion hertz
MHz = megahertz, or 1 million hertz
kHz = kilohertz, or 1,000 hertz

Cell phones use frequencies between the VHF and UHF ranges.

HOW CAN RADIO WAVES TRANSMIT OUR SOUNDS?

Since the range of human hearing (and much of our noise making) is much lower in frequency than the frequency of radio waves, the answer lies in *modulating*, or slightly changing, one wave—the high-frequency *carrier wave*—to represent another wave, the low-frequency *signal wave*.

When we talk about audible frequencies, we mean waves of air. When we talk about visible frequencies, or radio frequencies, we mean electromagnetic waves.

COMMUNICATION USING RADIO WAVES INVOLVES CHANGING THE WAVE

When you listen to a radio, you experience the changing of waves. Common ways to change waves are changing the frequency or changing the wavelength.

Analog systems use data that are continuous. Analog systems are built to be able to interpret all changes in the wave. When you listen to an ordinary radio, the music and talking you hear is the result of an analog system. The radio receives all the changes in the radio waves and uses those changes to reproduce the sounds.

Digital systems use discrete data—data that are grouped and represented by numbers. They interpret changes to the signal wave as 1s and 0s. The 1s and 0s are transmitted and on the receiving end, they are used to reproduce the signal wave. While some information can be lost, digital systems use *error correction* so that if a part of the signal is sent or received incorrectly, the receiver can still figure out the signal.

Waves can be changed in other ways, and cell phones often change waves in more than one way.

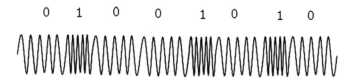

Frequency modulation: This sequence is 01001010. Here the amplitude of the wave doesn't change, but the frequency does. The part of the wave that has a higher frequency indicates a 1; the lower frequency indicates a 0.

Amplitude modulation: This is the same sequence: 01001010. Here the frequency of the wave doesn't change, but the amplitude does. The larger amplitude indicates a 1; the lower amplitude indicates a 0.

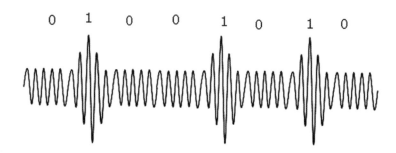

So why does a CD sound better than the radio? One reason has to do with transmission of the wave. A CD can send its digital signal with very little *interference*—interruption of the wave—to the speakers that reproduce the sound. Radio waves must travel through the air to reach your radio before the sound is reproduced. Another reason has to do with *sampling*, or how accurately the 1s and 0s reproduce the original wave. A higher sampling of the original wave reproduces the sound more accurately, but it requires much more information to be transmitted. We won't say much more about sampling here. The sampling, coding, and decoding of a wave is a complex and interesting topic that requires a book of its own.

Analog systems read data as continuous streams. When your eye and brain connect the shaded boxes to interpret the word *Bob* in our chapter 3 example, they are acting as an analog system. Digital systems break everything into 1s and 0s—the shaded or unshaded boxes for the computer in our example.

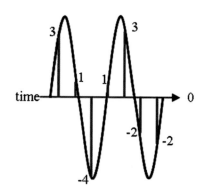

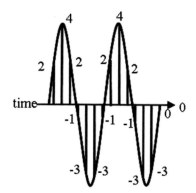

Sampling is a way to represent a wave as 1s and 0s. Each protocol has its own method. A simple method is shown here: the amplitude of the wave is measured and recorded as a number at different time intervals. The smaller the time intervals, the more frequent the measurement and the more accurate the representation of the wave.

THE CHANNEL IS MODULATED TO CONVEY THE SIGNAL

The radio-frequency (carrier) wave is also called a *channel*. Small changes to this carrier wave can be made to send the signal. Sound is converted by a *diaphragm* into an electrical signal. This signal is then used to change the carrier signal. When the modulated radio wave is received, the carrier frequency is already known, so the signal changes to it can be detected.

To imagine this better, return to the campsite. Suppose instead of using flashes of colored light, you and your friend use a steady red light because the camp counselors are less likely to wake up from it. You could change the frequency of the red light by going from a bright red to a deep red. The frequency doesn't change enough to change the reddish color you and your friend see; it's still red and won't interfere with your yellow- or blue-light neighbors. But you can distinguish a pattern in the bright- to deep-red range.

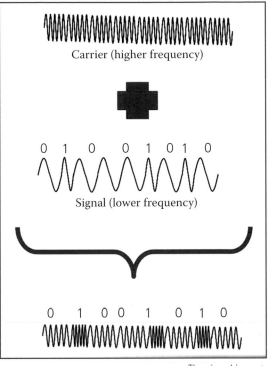

Carrier (higher frequency)

0 1 0 0 1 0 1 0

Signal (lower frequency)

0 1 0 0 1 0 1 0

The signal is sent in the lower-frequency wave. The pattern of change in the signal is transmitted in the higher-frequency carrier wave.

(This diagram shows just frequency modulation. Remember that cell phones use many ways to alter the wave.)

Radio Transmission: Putting the "Cell" into "Cell Phone"

IT TAKES A SYSTEM TO MAKE A CALL

As a wave gets farther from the source transmitting it, the small changes that represent the signal become harder to detect. You can see this effect with your flashlight in a dark room. If you shine the flashlight close to the wall, the circle of light looks bright. As you move the flashlight away from the wall, the circle of light gets bigger but fainter. The same light covers a larger area, so in each part of that area, the light isn't as bright. So how can your cell phone use radio waves to make a call across many miles or across the country? Do calls to more distant places take more battery power? The answer is no, because the cell phone that you carry is only a part of a larger network. The cell phone uses a system of cell towers and *mobile telephone switching offices* (MTSOs) to repeat and carry the signal.

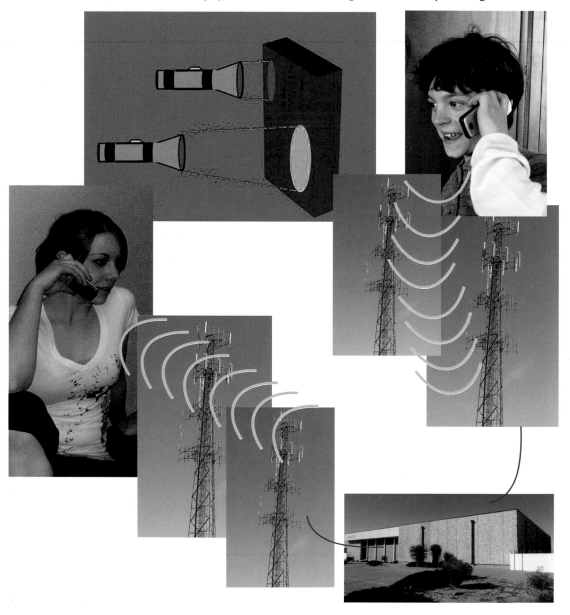

CELL TOWERS SEND AND RECEIVE RADIO SIGNALS

They are arranged in a honeycomb pattern that carves up the area they cover—the coverage area—into cells. Each tower broadcasts a different channel. Depending on where a phone is, one of the channels will always be stronger and the phone will use that channel, and that tower, for communication. A cell phone using a different channel is like the campers each using a different-colored filter for signaling.

The honeycomb pattern is the most efficient pattern to cover an area with the fewest towers.

THE LARGER SYSTEM ROUTES CALLS

The cell towers connect to the MTSOs. The MTSOs make the switching decisions for the cellular system by routing calls to other cell towers if the other cell phone is in the same, or *local*, area. The MTSO system also connects to the *public switched telephone network* (PSTN) and routes calls to and from it if one of the phones in a call is a landline phone (connected by wire to the PSTN).

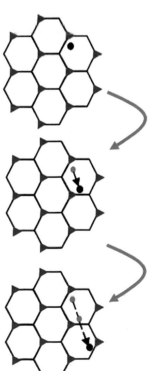

This diagram shows a simplified version of how you would see cell towers (blue triangles) cover an area (hexagonal outlines) if you flew over them in a plane. Each tower broadcasts a different carrier frequency. So each cell has three different channels available within it. When you turn on your phone, it detects the strongest channel and uses it (red triangles). This means that your phone communicates with that cell tower. When you move, your cell phone may detect another channel that's stronger than the one it's currently using. In that case, the towers use their own protocol to *hand over*, or change, the tower that's communicating with your phone. That way, your phone communicates with only one tower at any time.

PROTOCOLS WRITTEN INTO YOUR CELL PHONE CONTROL ITS COMMUNICATION

Turning your cell phone on seems like a simple thing to do, but by now you know that a lot of information gets sent between your phone and the cell towers before you ever receive a call or punch in a number. Your cell phone uses a protocol, written into its software in the phone's chip, to let the nearest cell tower know that it's there and that it will use that tower to communicate. Because the protocols are already in the software, you don't need to learn them—the system knows where to send any calls or messages for that phone without you having to do anything. As you move around, a different channel may get stronger and the channel your phone is currently using will get weaker; protocols control how the towers hand off control of your phone's communications from one to the other.

TAKING TURNS ALLOWS MULTIPLE USERS TO USE THE SAME FREQUENCY

Let's return to the campsite again. Now a new camp counselor has joined the group. This counselor is a very light sleeper and is sensitive not only to noise but also to lots of light. The concern of the campers is that now all of you can't send signals at the same time because this new counselor might wake up from the brightness of all the flashlights. How do you still manage to communicate after dark? The answer is taking turns with time instead of using colored light. Here's how it works: you assume that if only one-eighth of you flash your lights in your tents, the new counselor won't wake up. Since only one of a pair is using his or her flashlight at a time, the campers can divide into four equal-sized groups. The first group has their conversations during the first 15 minutes of every hour, the second group has their turn during the second 15 minutes, and so on. Now you and your friend know when you can have your conversation, even with the fewer flashlights working at a time.

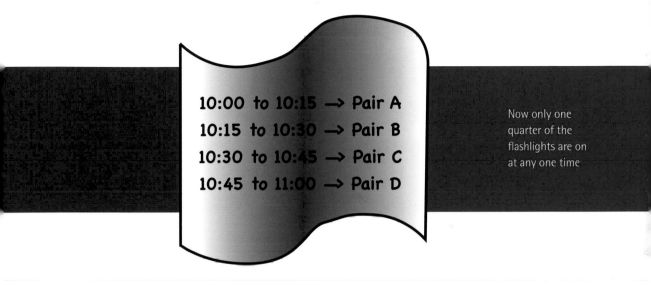

10:00 to 10:15 → Pair A
10:15 to 10:30 → Pair B
10:30 to 10:45 → Pair C
10:45 to 11:00 → Pair D

Now only one quarter of the flashlights are on at any one time

TDMA IS USED TO HANDLE LOTS OF CALLS

This time-sharing concept, called *time division multiple access* (TDMA), is used to handle increasing numbers of cell phone users. With real-life cell phones, it's not that we're scared of waking the new camp counselor—we don't want to overwhelm the cell tower with too many call signals at the same time. Of course, the time divisions for cell phones are much smaller than the 15 minutes used in the campsite. They're so small that you don't notice your call is "broken" in time.

THERE ARE OTHER WAYS TO ALLOW MORE PEOPLE TO TALK AT THE SAME TIME

TDMA isn't the only way to share cell towers. The use of multiple channels within a cell is an example of *frequency division multiple access* (FDMA); it's similar to using the different colors of red with a flashlight signal. In *code division multiple access* (CDMA), different protocols are used. This is similar to the campers using their own codes to communicate.

ONCE YOUR CELL PHONE RECEIVES THE SERIES OF 1S AND 0S, HOW DOES IT ACTUALLY INTERPRET THE MESSAGE?

The cell phone's protocols, which are written in its software, tell it how to respond when it's receiving an incoming call, a text message, or an Internet page. The pattern of 1s and 0s in the beginning of the message sequence tells the phone's computer what kind of message it is. So, for example, a message that starts out with a 10001001 may be an incoming call, which the phone reproduces using the rest of the message and its speaker. A message that starts with a 00100101 may be a text message, which it would display. You don't need to interpret the 1s and 0s; you don't even need to understand the system to make your call or send your text message. All you need to do is know how to punch in another phone number and your phone and its system does the rest!

In this imaginary protocol, the first eight digits of each signal (yellow) tell whether the incoming message is a phone call or a text message. The rest of the signal (blue) can be the message or the call or may even contain other information, like when the signal was sent.

CHAPTER 7

Batteries Are Gas Tanks Full of Electrons

Your cell phone needs to have its own portable power source to be truly mobile. Batteries provide mobile power in a light, sturdy package that's also rechargeable. Here's the fascinating story behind a key improvement in technology for cell phones.

What Is a Battery Anyway?

YOUR CELL PHONE MUST HAVE A SOURCE OF ELECTRICITY

A cell phone uses electricity to send and receive radio signals with its antenna, to power its display, and to do the many other things that it does. In addition, this electricity source must be mobile, and for convenience, it must last a fairly long time. The electrical energy that a battery—a *non-rechargeable* battery—produces comes from a chemical reaction. (We'll talk about rechargeable batteries later in this chapter.) A battery is actually an energy converter: it converts chemical energy into electrical energy

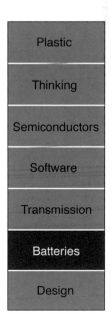

Plastic

Thinking

Semiconductors

Software

Transmission

Batteries

Design

(electricity). How does a chemical reaction produce electricity, or a flow of electrons? You need what is called an *electrochemical cell*, in which a chemical reaction uses and releases electrons. To understand how an electrochemical cell works, we'll need to start by briefly reviewing our description of the atom.

There are many different kinds of energy—chemical and electrical energy are just two kinds. Heat, sound, and kinetic (movement) energy are other kinds. Energy can be transformed from one kind into another, but when it is transformed, some of it doesn't go where we want it to. We call this an *energy loss* (even though energy can't be created or destroyed).

Close-up of an atom: the electrons circling the nucleus are negatively charged. The protons in the nucleus are positively charged, and the neutrons are electrically neutral. Atoms of the same element all have the same number of protons, but they can have differing numbers of neutrons (isotopes) or electrons (ions).

Negative electrons

Positive nucleus

This is a model of an atom. Remember, "All models are wrong, but some are useful."

THE ATOM IS THE SMALLEST PART OF AN ELEMENT THAT STILL HAS ALL THE PROPERTIES OF THAT ELEMENT

(Remember what we said in chapters 2 and 4?) Remember also that atoms themselves have even smaller parts: electrons, protons, and neutrons. Protons and electrons have equal and opposite electrical charges; neutrons have no charge. So atoms are electrically neutral. They have no electric charge since they have an equal number of protons and electrons—positive and negative charges.

Protons and neutrons are composed of even smaller particles, called subparticles. There are 19 subparticles in all, but we won't cover them here. (Electrons themselves are one of the kinds of subparticles.)

THE BASIC STRUCTURE OF ALL ATOMS

Neutrons and protons are in the center, or *nucleus*, of the atom, while the electrons *orbit* (follow a path around) the nucleus. Atoms of different elements have a different number of protons in the nucleus; this is what makes one element different from another. So each and every hydrogen atom has just one proton in the nucleus; all oxygen atoms have eight protons, and so on. When atoms of the same element have a different number of neutrons, these atoms are isotopes of that element.

CHEMICAL REACTIONS COMBINE ATOMS

In chapter 2, we said that molecules are made up of atoms. But what does combining atoms into molecules mean? Molecules are made by chemical reactions. In chemical reactions, the nuclei of the atoms aren't affected at all—the electrons orbiting the nuclei change their orbit.

ONLY VALENCE ELECTRONS ARE INVOLVED IN CHEMICAL REACTIONS

Since electrons never collide with each other as they orbit the nucleus, their orbits get complicated when there are more than one electron. So each electron has its own path at a certain distance from the nucleus of the atom. You can think of the atom as having electrons orbiting in groups and each group orbits at a different distance from the nucleus. The orbits form *shells*, or layers, around the nucleus of the atom, much like the layers around the core of an onion, only they aren't that close to each other. Some shells are very close to the nucleus; others are farther away. The farthest shell from the nucleus of the atom is called the *valence shell*. The electron (or electrons) in the valence shell changes its orbit in chemical reactions. When the valence electron's orbit is changed, the atoms are said to be *chemically bonded*, or attached chemically. Molecules are collections of atoms that are chemically bonded.

Want to see a chemical reaction? Add a quarter cup of vinegar to two tablespoons of baking soda. (Get help from an adult and do it in a container in the sink for easier cleanup!)

Nuclear Reactions

If we combine the nucleus of one atom with that of another, we get a single nucleus that has a different number of protons than either of the reacting atoms—*this is a different atom!* It's not a molecule because it has only one nucleus. We can't break this new atom apart chemically to get back the original reacting atoms.

When the nuclei of atoms are combined, the process is called *fusion*. When the nucleus of an atom is split apart—that is, the protons in the nucleus are divided into two (or more) nuclei—the process is called *fission*. Fusion and fission are called nuclear reactions because the nucleus of the atom is changed.

Nuclear reactions release very large amounts of energy. This is why nuclear energy reactors are used. Controlling the large amount of energy released in nuclear reactions can be difficult and dangerous—that's why nuclear energy reactors in the United States (and in most places) are very heavily regulated.

In some shells, the electrons orbit along oddly shaped paths. We won't get into those details here. Linus Pauling, Neils Bohr, and many others have worked on formulas, equations, and rules to describe these paths, but we still can't say for certain where to find an electron at any particular moment!

The electrons of an atom orbit the nucleus. Each electron orbit has its own shape.

Although some of the orbits are very complicated, the electrons' paths look like clouds around the nucleus from our point of view.

When many electrons orbit the nucleus, their cloud-like orbits group into shells around the nucleus.

(The colors in these diagrams are used only to distinguish between different electron orbits. Electron orbits don't really have colors that we can see.)

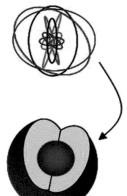

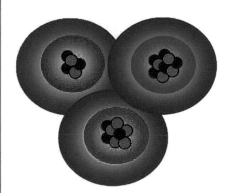

In the imaginary molecule shown at left, the nuclei (shown as red protons and yellow neutrons) don't combine and the inner electrons (orange, green, and blue around the nuclei) remain in their orbits.

Only the valence electrons (purple) now orbit around all of the nuclei, binding—but not combining—the nuclei together. Because of this change in electron orbit, the atoms are now a molecule and the molecule has different properties from each of the separate atoms.

THE STUDY OF CHEMICAL REACTIONS IS THE STUDY OF HOW VALENCE ELECTRONS CHANGE THEIR ORBITS

This means understanding which elements have atoms that are likely to chemically react, which are unlikely to chemically react, and how much energy a reaction will release or require when it takes place.

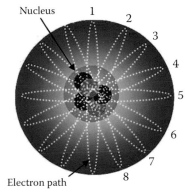

NOBLE GASES DON'T REACT CHEMICALLY

Elements whose atoms have exactly eight electrons in their valence shell (right) are *inert*. That is, they don't react with other elements because the amount of energy needed to make them change their orbits is very large. Their valence shell is said to be *stable*. These elements, called *noble gases*, are helium, neon, argon, krypton, xenon, and radon. They are all gases, not liquids or solids, because they don't even combine with each other!

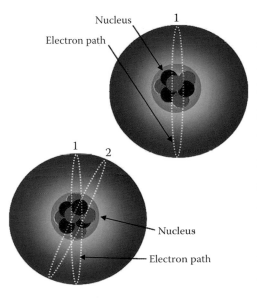

ALKALI AND ALKALINE METALS ARE VERY REACTIVE

Some elements have atoms with only one or two electrons in their valence shell. They are called the *alkali metals* if they have one valence electron (left, top) and *alkaline metals* if they have two valence electrons (left, bottom). These elements are very reactive, some extremely so. For example, when *potassium* (an alkali metal) comes into contact with water, potassium hydroxide and hydrogen gas are produced in an explosive reaction. This reaction releases a lot of energy and needs no help to get started. That's why the pure element potassium doesn't exist in nature and must be kept in oil (and away from water) at all times in a lab. You can think of alkali and alkaline metals as having atoms that like to release their electron(s) from their valence shells. When they do so, energy is also released.

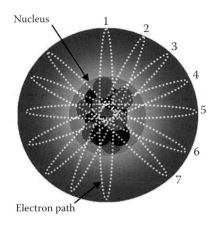

Nucleus

1 2 3 4 5 6 7

Electron path

The element astatine is also a halogen in the periodic table of the elements, but it doesn't exist in nature. It can be created in a nuclear reactor.

HALOGENS ARE ALSO VERY REACTIVE

Elements whose atoms have seven valence electrons are also very reactive. You can think of these atoms as liking to add an electron in their valence shell in order to create a stable arrangement of electrons. Energy is released when these elements add the eighth electron in their valence shell. These elements are called *halogens*. Like the alkali metals, individual halogen atoms don't occur in nature and they react vigorously with other elements. The halogens are chlorine, fluorine, bromine, and iodine.

Getting Electricity from Chemistry

A BATTERY IS AN ELECTROCHEMICAL CELL

To create a flow of electrons from a chemical reaction, we need:

- ✹ A chemical reaction that will release more energy than it needs, and
- ✹ A way to collect the electrons from it.

A chemical reaction that involves an electron transfer between the atoms usually releases more energy than it needs to get started and keep going. The particular elements and molecules involved, and how much of them are available to react, determine how much energy is released. To collect the electrons from a chemical reaction, we need a way to separate the *reactants*, or substances reacting in the chemical reaction, from the products of the reaction so that only the electrons flow between the substances. Such a setup is called an electrochemical cell. A battery is just a package for an electrochemical cell.

Think of a ball rolling down a hill. Stopping it takes much more energy than if it were rolling on a level road.

AN ELECTROCHEMICAL CELL HAS TWO ELECTRODES IN A SOLUTION

An *electrode* is a contact—a connection between two electrically conductive materials. The two electrodes in an electrochemical cell make a connection between the electrical circuit and the *solution*. This means that electrons flow through the electrodes to your cell phone and back to the solution. When the battery is *discharging*—supplying electricity—the electrically positive electrode is called the *cathode* and the electrically negative electrode is called the *anode*.

THE SOLUTION IS AN ELECTROLYTE AND ALLOWS IONS TO MOVE

The solution isn't just any old liquid. It's called an *electrolyte* and can conduct electricity. We mentioned earlier that some atoms prefer to release their valence electrons and others prefer to add them. When these atoms either release or add valence electrons to make a stable valence-electron arrangement, their electric charges are out of whack, or imbalanced, and they are called ions (remember ions from chapter 4?). An atom that has more electrons than protons is called a *negative ion* because it has more negative than positive charge. In the same way, an atom that has fewer electrons than protons is called a *positive ion* because it has less negative charge than positive. So potassium would create a positive ion and chlorine would create a negative ion. When ions are dissolved in water, they move freely among the water molecules to create an electrolyte.

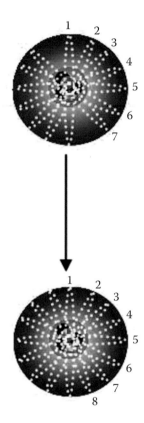

This potassium atom is electrically neutral because the number of electrons and protons is equal.

This potassium ion is a positive ion because it has fewer electrons than protons. (Shown here is the next-lower electron shell—the ion is smaller than the atom!)

This chlorine atom is electrically neutral because the number of electrons and protons is equal.

This chlorine ion is a negative ion because it has more electrons than protons.

THE ELECTROCHEMICAL CELL SEPARATES THE CHEMICAL REACTION INTO TWO PARTS

Each electrode in an electrochemical cell is made of one of the reactants in the chemical reaction that powers the battery. The electrolyte itself is also a reactant. So half of the reaction—a *half-reaction*—occurs at each contact. But for the battery to work, each half-reaction must occur naturally—that is, without needing any energy to start.

At the anode, the half-reaction releases electrons, and at the cathode, the half-reaction uses them. This makes a constant flow of electrons from the anode to the cathode through the electrical circuit. It's this constant flow of electrons—electricity—that powers the cell phone. The electron flow continues until one of the reactants runs out.

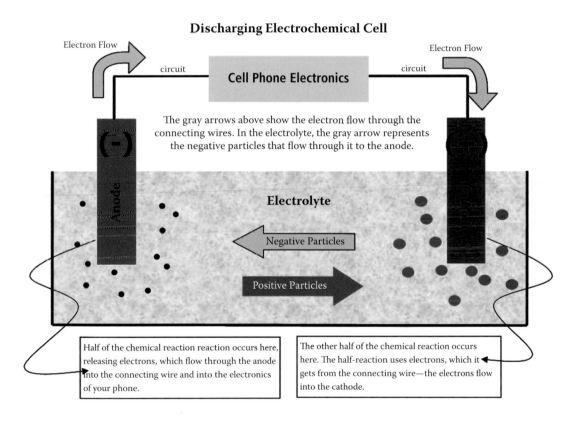

Discharging Electrochemical Cell

Electron Flow

Electron Flow

circuit

circuit

Cell Phone Electronics

The gray arrows above show the electron flow through the connecting wires. In the electrolyte, the gray arrow represents the negative particles that flow through it to the anode.

(–)

Anode

Electrolyte

Negative Particles

Positive Particles

Half of the chemical reaction reaction occurs here, releasing electrons, which flow through the anode into the connecting wire and into the electronics of your phone.

The other half of the chemical reaction occurs here. The half-reaction uses electrons, which it gets from the connecting wire—the electrons flow into the cathode.

In the electrochemical cell (above), the negative-particle flow balances the positive-particle flow. Negative particles can be ions or electrons; positive particles are ions. During discharge, the positive particles flow to the cathode and the negative particles flow to the anode. The half-reaction at the *cathode* requires positive particles; the half-reaction at the anode releases electrons which flow through the circuit to power the cell phone electronics. These flows are balanced as long as the half-reactions continue to occur at each electrode.

You may have heard of electrolytes before; they're often advertised in sports drinks. The electrolytes in sports drinks have the same characteristics as electrolytes in batteries: they allow the movement of charged particles. In your body, this charged-particle movement allows you to move, digest, and so on. In a battery, the charged-particle movement is converted to electricity.

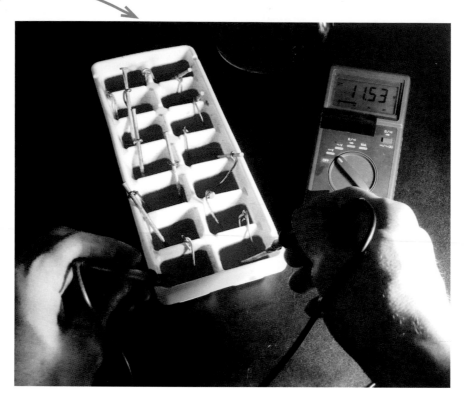

In fact, you can make a battery out of some sports drinks. The picture at right shows a battery that's made by pouring a sports drink into an ice tray and connecting each mold with an electrode made of galvanized steel or copper. This battery produced over 11 volts!

SO WHAT KIND OF CHEMICALS ARE USED IN A REAL BATTERY?

You may be familiar with alkaline batteries—they're everywhere. In this kind of battery, the electrolyte is potassium hydroxide dissolved in water—the potassium ions (positive) and hydroxide ions (negative) move separately in the water. The electrodes are made of powders: *zinc* powder for the anode and *manganese* dioxide powder for the cathode.

The ions in the potassium hydroxide solution can move fast. The hydroxide ions near the anode react with the zinc to create zinc oxide and water, releasing electrons, which flow into the circuit. At the cathode, the manganese dioxide and water use electrons from the circuit to react, forming manganese oxide and hydroxide ions. The hydroxide ions flowing through the potassium hydroxide solution—from the cathode to the anode—are balanced by the flow of electrons—electricity—through the electrical circuit. The flow stops when one of the electrodes is used up.

Alkaline Battery Electrochemical Cell

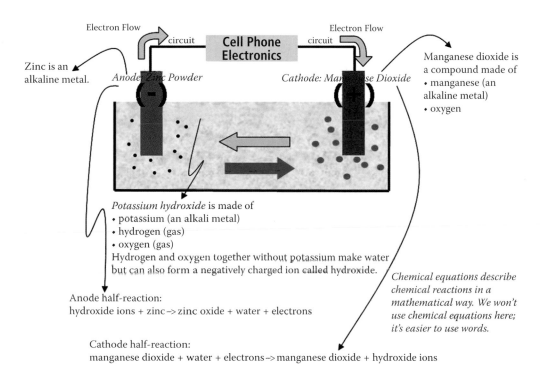

Electron Flow

circuit

Cell Phone Electronics

circuit

Electron Flow

Zinc is an alkaline metal.

Anode: Zinc Powder

Cathode: Manganese Dioxide

Manganese dioxide is a compound made of
• manganese (an alkaline metal)
• oxygen

Potassium hydroxide is made of
• potassium (an alkali metal)
• hydrogen (gas)
• oxygen (gas)
Hydrogen and oxygen together without potassium make water but can also form a negatively charged ion called hydroxide.

Chemical equations describe chemical reactions in a mathematical way. We won't use chemical equations here; it's easier to use words.

Anode half-reaction:
hydroxide ions + zinc –> zinc oxide + water + electrons

Cathode half-reaction:
manganese dioxide + water + electrons –> manganese dioxide + hydroxide ions

HOW ALKALINE BATTERIES HAVE IMPROVED

Since alkaline batteries were introduced in the 1960s, they have been improved in several ways. Better refining processes make the manganese for the manganese dioxide more pure. This means a faster half-reaction at the cathode. Adding carbon to the manganese dioxide further speeds the cathode reaction. Zinc oxide added to the zinc as well as a gel added to the potassium hydroxide slow the reaction at the anode. This makes the half-reactions balance better and prolongs the life of the battery.

How Can a Battery Be Rechargeable?

REVERSING THE CHEMICAL REACTION TAKES ENERGY . . . BUT HOW MUCH ENERGY IS THE KEY QUESTION

The cathodes of a lithium battery are made of a lithium oxide, like lithium cobalt oxide. The anodes are made of carbon. Lithium ions can insert themselves in both of these materials, much like grapes shoved into a stack of oranges

In lithium-polymer batteries, a polymer membrane (which insulates, remember?) acts as an electrolyte and separates the electrodes (anode and cathode) while letting the small lithium ions pass through it.

Now that we understand the basics of how a battery works, it's easy to say that making a battery rechargeable means we have to somehow reverse the chemical reaction. But this is like trying to get a ball to roll uphill: you can do it, but you have to keep adding energy, sometimes lots of it. So for a rechargeable battery, the question is: how much energy must be added to reverse the chemical reaction? Some products of chemical reactions don't easily revert to their original reactants. These products need more energy to reverse the reaction than the battery will release when it's discharged. So rechargeable batteries can use only certain kinds of electrodes and electrolytes. You can think of rechargeable batteries as a way to store electrons—like a gas tank stores gas.

MOVING IONS ALLOW LITHIUM BATTERIES TO CHARGE AND DISCHARGE

Lithium-ion and lithium-polymer batteries don't create and reverse chemical reactions. Instead, the lithium atoms change to lithium ions

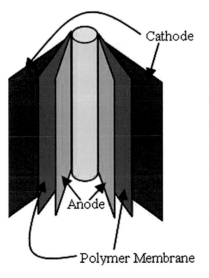

Cathode

Anode

Polymer Membrane

during discharge and change back from ions to atoms when the battery charges. The electrolyte in each of these kinds of batteries isn't involved in these changes—it just helps the ions to move.

NICKEL-METAL-HYDRIDE BATTERIES USE AN INTERESTING PROPERTY OF NICKEL

We said earlier in this chapter that only valence electrons are involved in chemical reactions. Well, some atoms can use more than just these electrons to form molecules—they can use electrons from the next-outermost shell as well. Nickel is one element that has this property: it can use either two or three electrons to form different molecules, and by going back and forth between these molecules, electrons can be used or released.

RECHARGEABLE BATTERIES HAVE MEMORY (AND OTHER) PROBLEMS

One problem with rechargeable batteries is memory. When a battery is discharged and recharged again and again, the materials can get degraded—used up—so that they can no longer react and reverse the reaction as well as when the battery was new. As a result, they may no longer provide as many electrons or for as long. In the past, this was a problem especially with nickel-cadmium (NiCad) batteries. Modern lithium and nickel-metal-hydroxide batteries don't have this problem as much. Nickel-metal-hydride batteries are expensive because

Remember at the beginning of this chapter, we said that the atom is the smallest part of an element that has all of the properties of that element? So, saying that nickel has a particular property is the same as saying that each atom of nickel has that property.

At the anode, the nickel hydroxide and the hydroxide ions react chemically to form nickel oxyhydroxide (yes, it's a tongue twister!) and water, releasing electrons. The cathode is made of two chemically reacted metals. Water and one of the cathode metals use electrons to form a metal hydride (metal-hydrogen substance) and hydroxide ion. When the battery charges, these reactions are reversed.

The number of electrons an atom uses to form chemical bonds is called its *oxidation state*. Some atoms, like nickel and iron atoms, have more than one common oxidation state.

of the materials used in them: the metals in the metal electrode are difficult to purify. The problem with lithium batteries is that the electrolyte or the membrane separating the electrodes can break down. If this happens, the ions and electrons could come together (touch), releasing a lot of heat and possibly starting a fire.

In chapter 2 we said that rubber has memory—it snaps back to its original shape when pulled.

This has happened in the past and resulted in a recall of batteries

HOW HAVE BATTERIES CHANGED THE LOOK AND FEEL OF CELL PHONES?

Cell phones have come a very long way from the big, heavy bricks they started out as, and much of this progress is due to the batteries they use. Lightweight (weighing mere ounces), rechargeable, and offering a long battery (discharge) life, batteries have made cell phones far more convenient and portable than ever before. Cell phones can now be easily carried anywhere, anytime, and can be used just about anywhere, anytime. Even small children can pick them up and use them. This dramatic improvement has certainly been key to making cell phones as widespread as they are. But battery improvement isn't the whole story. Making all the parts—the shell, the software, the hardware, the transmission, and the battery—work together well is just as important, as we'll see in the next chapter.

Dr. Martin Cooper with the Motorola DynaTAC 8000X, one of the first cell phones, at the 2007 Computer X E21 Forum. Notice how big the phone is compared to his hand. Much of the weight and size of this phone was its battery

CHAPTER 8

Good Design: Bringing the Technologies Together

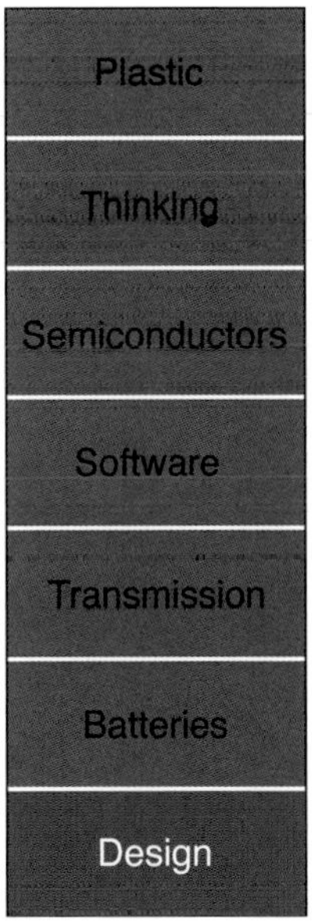

That the parts of a cell phone work isn't enough: *how* they work together is critical. Improvements to one part spill into others, making new designs possible. But to really take advantage of an excellent design, serious thought must go into how to build that design.

This chapter looks at how the parts of cell phones are brought together, who does this, and what these people (or you) might think of next. Putting cell phones together isn't pure science. It's the combination of art, science, and management that makes a great tool like a cell phone possible.

Plastic

Thinking

Semiconductors

Software

Transmission

Batteries

Design

EVERY ORGANIZATION THAT CREATES ITEMS TO SELL USES ITS OWN (SOMETIMES SECRET) DESIGN PROCESS

While we won't talk about a particular company or a particular process, we will talk about some things that every design process must address. How each organization actually accomplishes these tasks is as individual as the organizations themselves.

THE DESIGN PROCESS TYPICALLY STARTS WITH DESIGN CRITERIA—WHAT WILL BE CREATED?

This seems like a simple question, but it isn't. In fact, there are many questions to answer, such as:

- What will the product (cell phone) actually do?
- In what situations will it do these tasks?
- What will the product look, feel, sound, and *smell* like?
- How heavy or small will it be?
- Who will use it, where will they use it, and why?
- Who will buy it and why? (The user and the buyer can be different people!)
- How much will it cost to make?
- How much will the buyer be willing to pay for it?

THE DESIGN PROCESS FOCUSES EFFORT ON THE CRITICAL FEATURES

We'll talk more about who designers and engineers are and what they do later in this chapter.

The list of design criteria goes on and on and can include many more questions. To get answers to each question, *designers* and *engineers* can look at many kinds of information. Looking at past buying patterns tells them that somebody liked a particular product. Purchasing information can tell what features of a product people might especially like. Predicting what features people will want next can help to decide what will be important in the next design.

But a company can only build so many phones and so many different models of phone or the cost will be too high. The company has to find a way to focus the product's design. At some point, some design options must be ruled out in order to include others. The purpose of the design process is to define what features to include in a product and how to include them. These decisions are key to creating a product that buyers will find attractive enough to make them part with their money to buy it.

A GREAT PRODUCT DESIGN IS WORTH-LESS IF THE PRODUCT CAN'T BE BUILT

The product design is only one part of the overall design process. The manufacturing process must also be designed: it must reliably and repeatedly produce the product at a cost that matches the intended price at the mall *and* the intended profit of the company. If this seems simple, consider the difference between assembling 10 cell phones in a day and assembling 1,000 of them each day for days on end. It's much harder to assemble many phones repeatedly. What if a part doesn't fit with the others correctly? What if each part seems to work separately, but when you put the whole phone together, it can't make a call? The manufacturing design must deal with all these issues in great detail: how thick or thin a part must be, how much voltage should be detected at a certain circuit, and so on. If details like these aren't clear in the manufacturing design, making those 1,000 cell phones per day would be just about impossible.

Remember what we learned in previous chapters? In designing a phone, all of these parts must work well and work together!

- In chapter 2, we learned that the plastic used to make the shell of our phone is sturdy and can be formed into many different parts. Which parts in our new phone should be plastic? Which can't be?

- In chapters 3, 4, and 5, we learned about computers, hardware, and software and how they work together. The software must control all the components of the phone and display information for you. The more it does, the more hardware the phone needs.

- In chapter 6, we learned about the cell phone system and how it relays messages over large areas. Your cell phone's computer must code and decode the signals it receives and talk to the cell phone system.

- Finally, in chapter 7, we learned how batteries work. Your cell phone needs a battery that will supply electricity for all of the tasks it does and still be small and light.

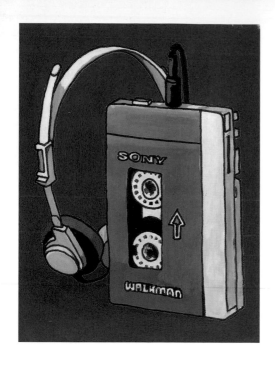

Who Does What?

WHO ARE THE PEOPLE WHO CREATE AND MANUFACTURE OUR PHONES?

We've talked until now about the decisions that need to be made to design a cell phone and the process to make it with. But processes don't make decisions—people do. Who are the people making these decisions and what special skills or training do they have?

ENGINEERS AND SCIENTISTS HAVE A BIG PART IN THE DESIGN PROCESS

Engineers and scientists often specialize in what they study through taking classes or on the job. There are many kinds of engineers and scientists who help to design a cell phone. We mentioned some of them earlier in this book.

- In chapter 2, we talked about plastics processing and how it affects properties. In chapter 7, we talked about getting electrical energy from chemical energy and how different materials help us to do that. These kinds of problems are what materials engineers and chemical engineers specialize in.

- In chapters 3 and 4, we talked about semiconductor chips, and in chapter 6, we talked about the cellular network. Engineers in many areas—electrical, computer science, materials science, chemical, mechanical, and industrial—work to design these parts of the phone.

- In chapter 5, we talked about designing, writing, and testing software. Software design includes a wide variety of roles, including software architects, software developers, software test engineers, and software project managers.

OTHER PEOPLE ARE NEEDED IN THE DESIGN PROCESS

Until now we've talked mostly about the people who are focused on the technical parts of design and manufacturing cell phones. But other people are just as important to bring a new cell phone to market:

- A project manager balances the scope, the schedule, and the budget of the project. He or she makes sure the right things (the scope) get done at an agreed-to pace (the schedule) and for an agreed-to amount of money (the budget). Without someone to keep things on track, a new design would almost never happen.

- A market research analyst gathers and makes sense of many different kinds of data and helps the team decide which phone features—like display size, keyboards instead of number pads, Internet access, and no antenna sticking out—are important.

- Attorneys and accountants help the design team watch for risk and profit while developing a product for sale. Most people don't think about them being in the design process—and they may not always play a big role—but their inputs can be very valuable.

ENTREPRENEURS HAVE A VISION AND THE DRIVE TO MAKE IT REALITY

An entrepreneur brings many things together: people to design and make a product, money and other resources to help the people, but most importantly, a belief that something new can be done. Entrepreneurs can seem like mad scientists: their visions seem far from what regular people think about. But entrepreneurs bring imagination to the market. Steven Jobs, Michael Dell,

Mad Scientist Wanted

Have you wondered whether mad scientists really exist? The answer is, that depends.

The mad in mad scientist comes from people thinking that some scientists are great at science but also crazy. Many people who accuse scientists of being mad have noticed that those scientists think differently from most people. But mad scientists are often very important *because* they think differently from most people. Without them, how can a new product be developed or a new model be imagined?

If no one had imagined connecting computers and having instant access to information around the world, the Internet wouldn't exist for us to download on our phones. If no one imagined using the bacteria that cause a sickness to create a cure, we would live in a much different, more dangerous world.

In the past, people believed that tomatoes were poisonous, that bleeding sick people would make them well, and that bathing would make people sick. Anyone who thought differently seemed crazy. Who seems crazy now?

So does it matter that scientists are sometimes shown in movies or on TV with bad haircuts and crazy expressions? Scientists themselves probably laugh at those shows because some scientists do look that way. But no one should avoid becoming a scientist or an engineer because of the mad scientist image. As a scientist or engineer, you can dress and act in any way you want—you just have to have a passion for learning how things work.

William Hewlett, and Kenneth Olsen are just a few entrepreneurs you may be familiar with. History is filled with them, and we seem to love having more.

Future Tools

WE'VE COME A LONG WAY . . .

. . . From a school bus on the way to overnight summer camp breaking down, through some of the history of *communication*, to the guts of our favorite electronic tool. What will happen next for cell phones? Although the future is hard to predict since technology improvements can happen so fast, here are some things that are already being tried or researched for future phones and services— and this list will change from the time we write it to the time it's published!

- ❦ *Wireless LAN* (local area network). This will make the Internet and other computer functions available on your phone.
- ❦ *Mobile TV.* To the thrill of advertisers and probably the horror of teachers and parents, television is becoming widely available on cell phones. One cell phone provider has TV in many markets already. Another company is working on short-episode programs that are custom-made for phones. These short programs can be downloaded more easily than long shows but will tie together to give a full story. Currently, viewers must pay for these downloads, but it's likely that advertisers will be willing to offer free downloads in exchange for commercial time.
- ❦ *Gaming.* Cell phone companies already provide some games on phones, but this area of phone service is being actively researched and expanded.

- *Credit or Charge Card.* This service that will also thrill some and scare others. In some countries, cell phones can already interact with vending machines and can be used as pre-paid cards or credit cards to purchase items. Someday the cell phone might replace the wallet completely. After all, a cell phone could conveniently carry all the information from a driver's license, credit cards, and bank accounts needed to get people through a normal day. But information security experts are worried about how to make this information difficult for a thief to steal.

- *Advertising.* The increased use of cell phones will tempt advertisers, and your phone service provider can increase its income by selling advertising—aimed directly at you through your phone. The same services that let you use your phone as a credit card might track what you buy and then use that information to send you special coupons.

- *More radio-based features.* The radio system that provides global positioning system (GPS) directions to a phone is now quite common. And local radio that provides Bluetooth has a similar story. What's next in radio features is less certain, but it's a good bet that your phone will send and receive additional, specialized signals to add these features.

THESE ARE JUST SOME OF THE IDEAS AROUND TODAY FOR IMPROVING YOUR CELL PHONE AND THE SERVICE IT PROVIDES TO YOU

A cell phone isn't just a tool used for conversation; it can be used for all kinds of communication. The possibilities for this small, portable, device seem endless, and anybody can come up with new ways to use it. Many of the ideas discussed here weren't even a possibility just a few years ago. New ideas inspire still newer ones. Only our imaginations limit us.

CHAPTER 9

Thank You for Turning OFF Your Cell Phone

Why do airline passengers have to turn off their cell phones? Does using a cell phone affect people's health? Why aren't cell phones allowed in some medical facilities? How private is a cell phone conversation? Just about everybody has these kinds of questions. They are the interesting result of the interaction between technology and people.

Why Would Cell Phone Use Be Prohibited?

THE CELL PHONE AS A DISASTER-RELIEF TOOL

Right after the attack on the World Trade Center in New York City on September 11, 2001, many cell towers had trouble keeping up with the usage. Many more people than usual were trying call friends and relatives with news. Getting news at a scary time like this is a huge relief, even if it takes a few tries to get through.

The cell phone has also become a part of many hikers' survival gear. Many have told stories of using their cell phones to help them get out of trouble. You know that people use their cell phones to call emergency services (police, ambulance) if they have—or see—car

wrecks. You may not know that relief workers and refugees in even the most remote disaster-hit regions use cell phones and mobile e-mail stations to arrange for help.

Although it pays to have a backup plan in case your phone doesn't work, these scenarios tell us a lot about just how much of our world is covered by cell phone signals. Whether it's a hiker calling search and rescue, a person calling for an ambulance after a car accident, or a disaster victim seeking food or medical aid, at times a cell phone is definitely just the right tool for the job. No doubt, cell phones have saved lives.

So why would cell phones be unwelcome at all? When is cell phone use impolite or even dangerous? Let's look at some of the places where cell phone use is restricted, knowing what we now know about how they work.

RADIO WAVES AND RADIO INTERFERENCE

Did you think about the places that don't allow cell phones while you read about the transmission of radio waves? Why could radio waves cause problems for other electrical devices? If you've been in a car with the radio playing and heard a moment of static while you passed beneath a power line, then you've experienced radio interference. The signal from the radio station's tower got interrupted because the power line disturbed the radio waves being transmitted. (The power line carries electricity, and so it has magnetic and electrical fields around it—it's an antenna!)

MEDICAL INSTRUMENTS USE RADIO WAVES

When you're listening to the radio, the information that gets lost because of the short interruption isn't likely to affect your health. But if you were depending on a medical device—for example, to monitor or stimulate someone's heart—radio interference at the wrong time might affect the patient's health. The interfering radio waves could even damage medical equipment, a very costly problem! If an instrument gave bad data, the patient might get the wrong medicine or might not get the right treatment. Because radio interference and medical equipment damage can be so costly, many medical offices are extra cautious about using cell phones. Medical equipment could be shielded to prevent interference and damage. But doing so would be expensive and temporary since cell phones are constantly changing.

AIRPLANE NAVIGATION SYSTEMS USE RADIO WAVES

Whether or not to ban using cell phones on airplanes is even more tricky. At one point, the Federal Aviation Administration (FAA) banned even studying the subject! The concern was that cell phones could interfere with the navigation systems of the plane, which also use radio waves. If pilots were misled by their own instruments on the plane, the results could be terrible. The FAA has recently lifted its ban on studying this subject.

Do Cell Phones Make You Sick?

ASSESSING RISK IS TRICKY

Will exposure to the radio waves that your cell phone sends and receives make you sick? What if you use your cell phone for a long time—years or decades? Will driving, walking, or living near a cell tower hurt your health? In assessing risk, it's difficult to decide how to compare the risks of different things. For example, is it more risky

Electromagnetic radiation has many different frequencies, energy levels, and sources. Humans and all life on the planet wouldn't exist without the sun, which is the biggest source of electromagnetic radiation in our solar system. We are also exposed to radiation from many sources in our universe, some of them very far away. On our own planet, elements such as radon, uranium, and plutonium are radioactive—meaning that the nuclei in their atoms split apart naturally and release radiation as these elements transform into other elements.

With so many sources of radiation, including electromagnetic radiation, around us, it's difficult to figure out what impact our cell phone use has on our health. In any case, less electromagnetic radiation is probably a good thing. So it's good news that cell phones these days use less and less energy when transmitting radio waves—cell towers are everywhere.

to ride in a plane or in a car? The answer depends on how much time you spend in each one, how much skill and experience the driver or pilot has, and what the road or air conditions are. Assessing health risks gets even more complicated. Each person is different and health can depend on height, weight, eating habits, exercise habits, family history, environment, mental capability, emotional outlook, and so on. It's difficult to decide which of these items to focus on. Also, when it comes to health, most experts would agree that everything, in some amount, is bad for you. The question is: where is that level?

IONIZING RADIATION CAN BE BAD FOR YOUR HEALTH

One thing we know is that *ionizing radiation*—radiation that knocks electrons away from the atoms it contacts—*can* be bad for your health. How bad depends on the length and level of exposure. Someone exposed to low levels of ionizing radiation probably won't get as sick, but high levels—like those near a nuclear explosion—can be deadly. (In Hiroshima, Japan, many people who survived the nuclear bomb explosion on August 6, 1945, died days later because they had been exposed to so much ionizing radiation.)

Do cell phones emit ionizing radiation and, if they do, how much? None has yet been detected. This isn't surprising since the amount of power modern phones use to transmit signals to the cell tower is so small. But some of the radio waves we come into contact with might be harmful. By using a cell phone, we increase our exposure to radio waves, a form of electromagnetic radiation (remember chapter 6?). In general, our exposure to radio waves has increased because cell phones and cellular systems now send radio waves everywhere. So no one knows if using your cell phone will make you sick someday. New studies might answer this question.

SOMETIMES, JUST USING A CELL PHONE IS RISKY

While we're thinking about risk, let's not forget that if someone is distracted by a cell phone call and has an accident in a car, on a bicycle, or just walking in the mall parking lot, the immediate risk of harm can be huge. Paying attention to what's going on around you is more important than using your phone to talk or text. Have you seen the bumper sticker that says "hang up and drive"? It's strange that a cell phone, so useful to call for help in an emergency, can be blamed as the distraction that *causes* an accident.

Privacy? What Privacy?

DO YOU KNOW HOW TO TURN OFF YOUR CELL PHONE?

Some security experts answer, "By taking out the battery." If that wasn't your answer, you might be surprised to learn that cell phones have been used to listen to conversations even though the people talking turned their phones *off*. Many cell phones can be remotely turned on with a radio signal. They can also be set up to act as a microphone and transmitter, *without the users' knowledge*, by a remote software change—software is

downloaded to the phone through a radio signal. When the government uses a suspected gangster's cell phone to listen to a conversation about an upcoming arms deal, *and they have permission from a court to do so*, this seems like a good thing. The fact that it can be done might make you worry about someone who doesn't have permission—or good and legal cause—doing the same.

A LANDLINE CALL IS MORE DIFFICULT TO INTERCEPT

Although this may sound too much like a spy movie to make you worry, there are corporate executives who routinely take the battery out of their phones. At certain government meetings (and private ones too), participants leave their cell phones outside the room. Some banks prohibit their employees from bringing cell phones inside their offices because those employees handle very sensitive financial information that could be photographed using a phone. Some companies even sell a detector that will warn of any signal coming from a phone that is intentionally or accidentally brought into such places. Why such precautions when (in the United States, at least) our privacy is legally protected? Because it's easier to intercept a cell phone call than a call on a landline, where someone has to physically connect to the phone system.

JAMMING A RADIO SIGNAL MAKES IT LOOK LIKE NO SIGNAL IS THERE

Could some kind of screen be set up to prevent phones from working in a theater or a top-secret meeting? This is called *jamming* the radio signal and can certainly be done. When we talked about radio transmission in chapter 6, we said that certain frequencies are used for cell phones. If you transmitted a signal that couldn't be interpreted, like a constant tone, on those frequencies, the phones within the

range of your jammer would seem to their users to have no signal. They wouldn't be able to communicate with the cell tower.

THE FACT THAT YOU CAN DO SOMETHING DOESN'T MAKE IT LEGAL

In the United States, no one but the government is allowed to jam cell phone signals, although the technology to do so exists. There are other ways to keep phone signals from where you don't want them. Conductive or magnetic material can be used in a room or building to create a screen against radio signals. So whether to jam or not is mostly a legal question, not a scientific or technological one.

(Groan) Someone's Phone Is Ringing

SO, WHAT ARE THE RULES FOR PROPER CELL PHONE CONDUCT?

When they're new, most technologies—like cars, telephones, music players, even call waiting—frequently cause new discussions, and sometimes arguments, about being courteous to others. Let's leave the really tricky situations to the etiquette experts, but some kinds of cell phone courtesy are easy to see.

In some situations, it's better to ignore your ringing phone, for example, when you are talking to someone face-to-face. Sometimes it's best to silence your phone because it might disturb others, like at

> So how do you decide if your cell phone is being an annoyance? Theaters, classrooms, and experts in etiquette have rules for you to follow. Other than that, just pay attention and use common sense.

a movie or in a classroom. Taking a picture of someone with your cell phone, even though they may not want you to, may seem like fun, but you might have to deal with legal or privacy problems—especially if you caught an embarrassing moment (did an image of a celebrity in an embarrassing situation just pop into your head?). And, of course, don't do anything illegal.

THE JOURNEY CONTINUES . . .

Your cell phone is an amazing little device that goes everywhere you do. It's a handy tool for keeping you in touch with your friends, family, and the world around you.

Now that you've learned some of the science and engineering that goes into making your phone work, we hope you see it in a different way—a way that makes you marvel at how it does what it does for you.

If You Want to Know More About Cell Phones . . .

Here are some places to look.

GENERAL CELL PHONE INFORMATION

www.fcc.gov/cgb/kidszone/faqs_cellphones.html
This site is on the Federal Communications Commission (FCC) website. It talks about different cell phone topics.

www.howstuffworks.com/cell-phone.htm
This site is a bit more advanced, but there are a lot of links to other topics.

www.ntia.doc.gov/osmhome/allochrt.pdf
Here's a diagram of all the frequency allocations managed by the FCC. Be warned—it will make you dizzy!

COMPUTERS, SOFTWARE, AND HISTORY

www.computerhistory.org/
The Computer History Museum website has a lot of information about early computing up to the present day.

www.intel.com/museum/onlineexhibits.htm
This page, on the Intel Corporation website, has more information on semiconductor processing.

people.ku.edu/%7Enkinners/LangList/Extras/langlist.htm
Here's one person's list of all the software languages out there. There are other sites that also do this.

BATTERIES
www.energizer.com/learning-center/Pages/how-batteries-work.aspx
This page, on the Energizer Holdings, Inc., website, has more information on how batteries work.

PLASTICS
pslc.ws/macrog/kidsmac/glossary.htm
Plastics glossary for kids.

GENERAL SCIENCE INFORMATION
web.mit.edu/course/3/3.091/www3/pt/
Here's one site with the periodic table of the elements. There are many others.

www.lbl.gov/MicroWorlds/ALSTool/EMSpec/EMSpec2.html
This page has a nice explanation in pictures of the electromagnetic spectrum.

Glossary

A

abstract: To summarize or draw meaning from a small amount of information.

acid: A *chemical substance* that easily gives up *hydrogen ions*. Acids vary from slightly to highly reactive and can be used to dissolve the exposed parts of many materials.

algorithm: A method or rule by which to do something.

alkali metal: An element that has one electron in its *valence* (outermost) *shell*.

alkaline metal: An element that has two electrons in its *valence* (outermost) *shell*.

amorphous: Having no pattern in arrangement throughout; disordered.

amplifier: Something that makes a signal bigger, louder, or more easily recognized.

amplitude: The distance between the peak and the valley of a wave.

analog: A mechanism or system which uses (or attempts to use) all the *data*. It doesn't try to break up the data in any way, as a *digital* system or mechanism would. You can think of this as *sampling* constantly.

anode: A connection between a *circuit* and *solution* (*electrolyte*) in an *electrochemical cell* at which the *half-reaction* that releases *electrons* takes place. *See also* electrode.

antenna: A rod or wire used to send or receive *electromagnetic waves*.

arsenic: An element that can be chemically combined with *gallium*, another element, to form *gallium arsenide*, a *semiconductor*.

assembler: A program that converts *source code* that a *programmer* writes into instructions that a *computer* understands.

assembly language: A set of words a *programmer* uses to write a *program* that a computer can understand. An *assembler* substitutes numbers for the words translating them into machine language.

atom: The smallest piece of an *element* that still has the *properties* of that element.

B

BASIC (Beginner's All-Purpose Symbolic Instruction Code): A *high-level programming language* that uses an *interpreter*.

biologist: A scientist who studies living things.

block copolymer: A *polymer* made of *monomers* in a repeating pattern. *See also* copolymer.

boot: To start up (a computer).

boron: An *element* used to *dope silicon* in silicon *semiconductor* manufacturing. It has fewer available *electrons* than the silicon so it changes the *voltage* at which silicon becomes *conductive*.

branched polymer structure: When the *monomers* that form a *polymer* are connected both in a line and with some that stick out from the main trunk.

bulk doping: To add an element or a substance throughout another substance. *See also* doping.

C

C: A *high-level programming language* that uses a *compiler* to create *object code* that the *computer executes*.

capacitor: An electrical device that can be filled with and store *electrons*. When the surrounding *circuit* has lower *voltage* than the capacitor, the capacitor *discharges*, or pours electrons back into the circuit.

carbon: An element found in hydrocarbons, which are a building block for polymers.

carrier wave: An electromagnetic wave having a *frequency* in a certain range that is used to *transmit signals*. It's changed slightly by the sender to encode the signal *wave*, and the changes are decoded by the receiver. *See also* channel.

cathode: A connection between a *circuit* and a *solution* (*electrolyte*) in an *electrochemical cell* at which the *half-reaction* that consumes *electrons* takes place. *See also* electrode.

CDMA. See *code division multiple access*.

cell: For cell phone networks, the smallest piece of a *coverage area*.

cell phone: A phone that uses *radio waves* and a *cellular network* to send and receive messages.

cellular network: A group of *cells, cell towers,* and *mobile telephone switching offices* that work together to *communicate*.

channel: An electromagnetic wave with a *frequency* in a certain range that is changed slightly to send *signals*.

charge: A measure of electricity, usually resulting from having an imbalance of electrons. Too many electrons is called negatively charged. Too few electrons is called positively charged.

chemical: *Elements* and *substances*—everything bigger than a subatomic particle is made up of some chemical or combination of chemicals.

chemical reaction: When two or more *atoms* combine to make a *molecule,* or when a molecule breaks apart into atoms or other molecules. *See also* reaction.

chemical bond: A connection between two or more *atoms* forming a *molecule.* The *valence electron orbits* are changed.

chemist: A *scientist* who studies *elements* and *substances.* Chemists are typically interested in the ways that elements and substances respond to each other and to changes in temperature, pressure, and *electric charge.*

chemistry: A field of science that studies how *elements* and *substances* respond to each other, *electric charge*, temperature, and pressure.

chip (semiconductor): A small piece of *silicon wafer* processed to contain many *integrated circuits.*

circuit: The path electric charge moves through. *See also* integrated circuit.

circuit board: A thin sandwich of conducting and insulating layers. Electronic devices are mounted to it and are connected to each other through it. A circuit board is usually green in color.

class: See *clean room class.*

clean room class: A measure of how clean a room is—the maximum number of particles smaller than 500 *nanometers* in a volume of air. There are many standards by which clean rooms are rated.

code: A set of instructions for a computer to follow; also, to create a set of instructions for a computer to follow. *See also* program.

code division multiple access: A method of letting many phone-tower or tower-tower pairs use the same *carrier waves* in a *cell*. Each pair uses a different *protocol* for its *two-way communication*.

communication: Sharing ideas or *data* by passing mutually understood data—words, numbers, or other things like specifically shaped *waves*.

compiler: A *program* that converts *source code* into *object code*. It collects and rearranges all the *instructions* to make the program run faster.

computer: A machine that performs mathematical calculations. Modern computers use mathematical calculations to perform much more advanced tasks like word processing, drawing, and communicating through the web.

computing power: Speed of calculation or speed of *execution* of *instructions*. Computing power is often measured in *millions of instructions per second*.

conductor: A *substance* or an *element* that allows easy flow of *electricity*.

copolymer: A plastic with molecules that combine different *monomers*. Monomers arranged in a repeating pattern are called *block copolymers*. Monomers arranged in a non-repeating pattern are called *random copolymers*.

coverage: The area in which a cell phone can send and receive signals.

crosslink: When *polymer molecules* stick to each other in one or more places.

crystal: Describes *substances* whose *atoms* are arranged in an three-dimensional repeating pattern. For any given atom within a crystal, the atoms nearby are in specific places. Common metals, table salt, and quartz are examples of *crystalline substances*.

crystalline: Refers to having a three-dimensional repeating pattern in the structure.

D

data: Information.

dense: Describes something that has a lot of mass for a given volume. Steel is dense compared to Play-Doh, so if you find the mass of a kitchen spoon (steel) and then form Play-Doh into an exact duplicate, the Play-Doh spoon will have a smaller mass.

deposit: To build up a solid material in a layer-by-layer way.

designer: A person who creates or improves concepts of how things should function or look so that people find them both attractive and practical.

diameter: The length of a straight line going through the middle of a circle from one edge to the other.

diaphragm: A thin, stiff piece of material designed to allow small back-and-forth motion—to create sound waves in the case of a speaker or pick up sound waves in the case of a microphone.

digital: A system or mechanism using data that is coded or processed in 1s and 0s.

discharge: Allowing something to leave. In the case of a capacitor or a battery, discharge allows the stored electrons to leave.

dopant: An *element* or elements added to a pure *semiconductor* to change its *properties*. *See also* dope.

dope: The process of adding an *element* or a *substance* to a pure substance to change its *properties*. In bulk doping, a *wafer* producer dopes a container of melted *silicon* with *phosphorus* before forming an *ingot*. In *semiconductor* processing, specific areas are doped.

download: To copy a *program* or data from one computer to another.

E

elastomer: A *polymer* in which only one out of every 100 molecules is *crosslinked*. This makes the polymer stretchy but able to bounce back to its original shape.

electric: Term for something designed to utilize the gathering or flow of *electrons*. The coffeemaker uses electron motion to heat water. The Hoover Dam uses water motion to create electron motion.

electricity: Physics term for gathering or flow of *electrons*. Static electricity collecting on a cloud *discharges* to another cloud or the earth through lightning.

electrochemical cell: A device that produces *electricity* from a chemical reaction.

electrode: An electrical contact device. In a battery, the electrode makes contact between the electrolyte and the external circuit.

electrolyte: A substance, usually liquid or solid, in which ions can move freely.

electromagnetic waves: Radiation that travels at the *speed of light*. Electromagnetic waves include radio waves, microwaves, heat, visible light, ultraviolet, x-rays, and gamma rays. They can travel in a *vacuum*.

electron: A subatomic particle with negative charge that orbits the nucleus.

electron shell: A grouping of *electron orbits*.

element: A *substance* that is made up only of itself. Gold is an element—there are individual gold *atoms*. Steel is a substance that isn't an element—it's made up of atoms of iron, *carbon*, and other elements.

energy loss: Energy that doesn't go or transform as intended. Energy isn't created or destroyed.

engineer: A person who applies scientific concepts to practical problems.

error correction: in digital systems, a method of repeating the ones and zeroes of the signal so that if a part of the signal is sent or received incorrectly, the message can still be deciphered.

etch: To eat away, dissolve, remove. In semiconductor manufacturing, plasma or acids are used to etch.

execute (an instruction): To carry out, perform.

F

fab: Commonly used slang for a semiconductor chip *fab*rication plant. A fab includes the clean room(s) and all of the equipment used to make the chips that are inside the clean room.

FDMA: See *frequency division multiple access.*

field: A space in which a force can be felt and can exert an influence. For example, a magnetic field is the space in which magnets can move magnetic objects.

filter: A device that separates desired from undesired. In the case of a filter for light, it lets through the desired color.

fission: Division into parts, but the word *fission* is often short for *nuclear* fission—the splitting of the atomic nucleus, causing the release of large amounts of energy. Current nuclear reactor *technology* uses fission, not *fusion.*

flash: Short for flash *memory*, which is a type of *persistent memory*. Flash memory transistors hold their state—1 or 0—whether power to the chip is on or off—until erased by a specific electric signal.

frequency: The number of waves that pass a certain point as the wave travels through space.

frequency division multiple access: A method of letting many phone-tower or tower-tower pairs use the same towers. Each pair uses a different *frequency* for Its *two-way communication.*

force: A physics term for an influence or influences over the motion of an object. The force of several people against a car bumper can cause the car to roll forward. The force of your hand can cause a rolling ball to stop.

fusion: Joining together. In *nuclear reactions*, two atomic *nuclei* join to form a different *element*. Fusion reactions power the sun and make hydrogen bombs powerful. In both cases, the energy comes from hydrogen nuclei joining to form helium.

G

gallium: A metallic *element*, used in making *gallium arsenide* for *semiconductors*.

gallium arsenide: A *semiconductor* material. Although less common than silicon, gallium arsenide is becoming a more widely used material in semiconductors because of its *computing power* advantages.

gate: See *transistor gate*, *logic gate*.

gate oxide: A very thin layer of insulating material that separates the *transistor gate* from the part that it controls. For a device *designer*, a gate oxide has to be thin enough to allow the *electric* influence from the gate above but not so thin that it allows electrons to move through the gate oxide, ruining the transistor.

glass: An *amorphous* solid material. The kinds of glasses we are most familiar with are made mostly of *silicon dioxide*.

gram: A unit of *mass* equal to the amount that one cubic centimeter of pure water weighs on earth.

H

half-reaction: A *chemical reaction* that either releases or uses *electrons* in an *electrochemical cell*. A half-reaction takes place at each *electrode*.

halogen: An *element* with eight *electrons* in its *valence shell*. It doesn't react with other elements.

hand over: The method, or *protocol*, that *cell* towers use to change which of them is the primary tower for a cell phone.

hardware: The part(s) of a *computer* that perform *logic* or *memory* functions but are built into *circuitry*.

hardwired: Built into *electric circuitry* and can't be changed by *software*. For example, the logic of the two light switches that control the same light in a room are hardwired by an electrician.

hertz: A measurement of *electromagnetic wave frequency*. It measures the number of *waves* per second and is abbreviated Hz. This unit was named after German physicist Heinrich R. Hertz.

high-level programming language: The set of words a *programmer* uses to instruct a *computer*—tell it how to perform tasks. *C* and *BASIC* are just two high-level software languages; there are hundreds of them.

hydrocarbon: A *substance* with *molecules* made of *carbon* and *hydrogen* chains.

hydrogen: The smallest, lightest *element*, typically consisting of one *electron* orbiting one *proton*.

hydrogen bond: The (*chemical*) connection between two *hydrogen* *atoms*. Hydrogen bonds can help to *crosslink* a *polymer*.

I

immobile charge: When *electrons* aren't able to move freely within a *substance*.

implant: A process for *doping* a *semiconductor wafer* by creating and shooting *ions* of the *dopant* into the wafer.

ingot: A piece of solid *material* formed into a long, skinny shape. In the case of a *silicon* ingot, a single *crystal* of *silicon* is pulled from a container of melted silicon into a long cylinder with tapering ends. Ingots of single crystal silicon are sawed to make *wafers*.

instruction: A direction or set of directions that tell a *computer* how to do something.

instructions per second: A measure of how fast a *computer* is—the more instructions a computer can *execute* per second, the faster it is.

insulator: Something that resists or prevents *electron* flow.

integrated circuit: A *circuit* or many circuits created on a single *chip*, giving the ability to perform complex *logic* functions quickly and with a small amount of *electricity*.

interference: A change caused in a *wave* as it travels through space. Interference may be caused by objects through which a wave can't pass or by another wave.

interpreter: A *program* that converts *source code* into *object* code one *instruction* at a time.

ion: An *atom* that is out of *electric* balance because it has too many or too few *electrons*.

ionizing radiation: *Waves* with enough energy to add or remove *electrons* from *atoms* that they strike.

isotope: Description of the number of *neutrons* an *element* has or doesn't have. The most common isotope of *hydrogen* has no neutrons, but there are also isotopes of hydrogen with one and two neutrons.

J

jam (a radio wave): Preventing *communication* by sending a strong, garbled *signal* at the *frequency* or frequencies used by the *radio* being jammed.

L

landline phone: A phone connected by wire (usually copper wire) to the phone system.

layer of abstraction: In a *model*, a level or stage that represents many things having the same structure or function. For example,

plasma: A type of *substance* considered to be different than liquid, solid, or gas. Plasma is gas-like but has a lot more energy, which lets its *ions* and *electrons* move freely. The electron and ion motion makes plasma highly *reactive*, which can be useful. Fire is a kind of plasma you're familiar with.

plastic: An *amorphous polymer* that can be molded into various shapes.

polycrystalline: Having more than one *crystal* in its structure.

polymer: A large *molecule* made primarily of *carbon* and *hydrogen*, organized into strands. Polymers can have many different sizes and structures and can also include other *atoms* in their structures.

polysilicon: *Polycrystalline silicon*—silicon having more than one *crystal*.

polystyrene: A *polymer* made of *carbon* and *hydrogen* (only).

positive ion: An *atom* with one or more *electrons* missing. It has a positive *charge*. *See also* ion.

potassium: A very reactive metallic *element*.

program: A set of instructions for a *computer* to follow; also, to create a set of instructions for a computer to follow. *See also* code.

property: A characteristic, quality, or trait.

protocol: A set of conventions, or rules, for *communication*.

proton: An *electrically* positive *particle* usually found in the *nucleus* of an *atom*. There is at least one proton in the nucleus of every atom.

PSTN: See *public switched telephone network*.

public switched telephone network: The landline phone system.

R

radio: Refers to a set of *electromagnetic waves* of certain *frequencies*. *Radio* can also refer to a device that sends radio waves—by *modulating* them—or receives radio waves—by interpreting the modulation of the waves.

RAM: See *random-access memory*.

random-access memory: *Volatile memory* that is used to store *data* to be processed or to store the results of processing. The *transistors* change their state to store data or the results.

random copolymer: A *polymer* made of *monomers* in a non-repeating pattern. *See also* copolymer.

react: To change or transform. In a *chemical reaction*, only the *electrons* are involved. In a *nuclear reaction*, the *nuclei* of the *atoms* are involved.

reactant: One of the *substances* that undergoes a *reaction*.

reaction: The act or process of change or transformation.

read-only memory: *Memory* designed to be easily retrieved. It's usually *persistent* memory.

rechargeable: A battery that can take in as well as supply *electricity*. The *chemical reaction* in a rechargeable battery is designed to be reversed easily.

refine: To make purer, or more usable.

ROM: See *read-only memory*.

S

sample: To re-create a signal by taking measurements of that signal at regular spaces in time. The shorter the spaces in time, the better the signal can be re-created.

scientific method: A set of steps used to discipline the discovery process in science.

scientific process: See *scientific method*.

scientist: A person who studies science, or how the world works.

score: To scratch precisely.

semiconductor: A material that isn't as conductive as a conductor (metal) but isn't as insulating as an insulator. It acts as an insulator at some voltages but at higher voltages is nearly as conductive as a metal.

service area: The set of *cells* in which a company's phones can send and receive signals.

shell: See *electron shell*.

signal: A message, instruction, or information that is transmitted from sender to receiver.

signal wave: The wave that contains or encodes the *signal*.

silica: See *silicon dioxide*.

silicon: An element. Silicon is a *semiconductor*.

silicon dioxide: A *substance* whose *molecules* are one *silicon atom* connected to two *oxygen* atoms. Silicon dioxide is the basic building block for glass (an *amorphous* solid) and quartz (a *crystalline* solid) among other things.

software: Refers to the instructions that a *computer* uses to carry out a task. Software has many levels, or *layers of abstraction*, between the *programmer* who creates the instructions and the *transistors* that carry them out.

solution: A mixture of solids, liquids, or gases in which each of the substances is evenly spread within the other(s). A jar full of air on your kitchen counter contains a solution of mostly nitrogen and oxygen with small amounts of many other gases.

source code: The set of instructions that a *programmer* creates in a *high-level software language*. See also *code*.

speed of light: An important constant in physics, light travels at about 186,000 miles in a second (a little less than 300 million meters in a second). All electromagnetic waves (discussed in chapter 6) travel at the same speed.

stable (valence shell): Refers to an *atom* or *ion* that doesn't *chemically react* with other atoms or ions because it has eight electrons in its valence shell. The *helium atom* is an exception—it doesn't react with other atoms, but has only two electrons in its valence shell.

state: Logic term for the condition of something. A light switch might be up or down. A lightbulb might be on or off. In these examples, up, down, on, and off are states.

store: To keep for later use. In *computers*, *transistors* are used to store *data*.

substance: A material that has the same properties throughout. It can be an element or a combination of elements.

T

TDMA: See *time division multiple access.*

technology: A way of completing a practical task or applying knowledge to solve a problem. A *cell phone* is more recently developed *communication* technology than the telegraph.

time division multiple access: A method of letting many phone-tower or tower-tower pairs use the same *carrier waves* in a *cell*. Each pair uses a different time slot for its *two-way communication*. These time slots are spaced so tightly that you don't notice them.

transistor: An electrical device that can act as a switch or an amplifier.

transistor gate: The part of a transistor that regulates the flow of electrons.

transmit: To send a message or a signal from one place to another.

two-way communication: *Signals* or messages in which a reply is expected. Two-way communication requires the use of *protocols*, or rules, to determine when signals should be sent and received. *See also* communication.

V

vacuum: A perfect vacuum contains nothing. A perfect vacuum is nearly impossible to create or find. The fewer atoms or molecules in a volume, the closer that volume is to a perfect vacuum.

vacuum tube: A container—usually a glass bulb—that holds an *electric* device sealed in a *vacuum* or a non-reactive gas.

valence shell: The outermost *electron shell* of an *atom* or *ion*.

volatile memory: Memory that doesn't keep *data* without electric power. When you turn off the phone or the battery dies, the data stored in its volatile memory disappears.

voltage: A measure of electric *charge* pressure. Negatively charged *electrons* collected in one place cause increased negative voltage.

vulcanization: A process for hardening and toughening rubber using sulfur and heat. Named for the Roman god of fire and metalworking, Vulcan.

W

wafer: A thin piece of *semiconductor* material that is sawed from an *ingot*—kind of like a slice of bread.

wave: A disturbance that transfers energy. Some waves, like those in the ocean or sound waves in air, need to travel through a substance. *Electromagnetic* waves can travel through a *vacuum*.

wavelength: Distance from a point on one wave to the identical point on the next. For many people it's easiest to think of measuring this from one peak to the next.

web browser: A *computer program* or set of programs designed to help find sites on the Internet. It downloads files from the Internet and displays them on the computer you're using.

weight: The force of gravity on a *mass*.

Z

zinc: A metallic *element,* usually used in *alkaline* batteries.

zone: For cell phone networks, the smallest piece of a *coverage area.* The same thing as a *cell.*

Illustration Credits

Page vi–vii   Cell phones courtesy of iStock

Page 7   Semaphore boy, Illustration by Cedra Wood

Page 9   Dr. Alexander Graham Bell, Illustration by Cedra Wood

Page 11   Server Farm, Illustration by Cedra Wood

Page 12   Map of NM with cells, Illustration by Cedra Wood

Page 24   Injection Molder, Illustration by Cedra Wood

Page 24   Polymer Pellets, Illustration by Cedra Wood

Page 32   Optical Illusions, Illustration by Cedra Wood

Page 37   Room with Switches, Illustration by Cedra Wood

Page 39   Alredeor and Por Lines, Illustration by Cedra Wood

Page 40   Bridge Out, Illustration by Cedra Wood

Page 40   Rockslide, Illustration by Cedra Wood

Page 44   John Von Neumann, Illustration by Cedra Wood

Page 44   Alan Turing, Illustration by Cedra Wood

Page 44   Woman Loading Punchcards, Photo courtesy of Computer History Museum

Page 50   DEC PDP-1 Minicomputer, Photo courtesy of Computer History Museum

Page 57   Tilted Lemonade with Valve, Illustration by Cedra Wood

Page 61   Silicon wafers, Illustration by Cedra Wood

Page 63   Ion Implantation Orange-Spy Model, Illustration by Cedra Wood

Page 77   Dog Ate My Homework, Illustration by Cedra Wood
Page 81   How RAM Works, Illustration by Cedra Wood
Page 81   RAM Capacitor Detail, Illustration by Cedra Wood
Page 82   RAM Array Storage, Illustration by Cedra Wood
Page 82   RAM Array Retrieval, Illustration by Cedra Wood
Page 83   RAM Capacitor Detail, Illustration by Cedra Wood
Page 84   Personal Computer, Illustration by Cedra Wood
Page 120   Pushing Ball on Incline, Illustration by Cedra Wood
Page 123   Dr. Martin Cooper, Public Domain Photo
Page 130   Sony Walkman, Illustration by Cedra Wood
Page 139   Listening Through Glass, Illustration by Cedra Wood
Page 141   Eavesdropping Over Newspaper, Illustration by Cedra Wood
Page 142   Annoyance in the Theatre, Illustration by Cedra Wood

All other images by the authors.

CDMA, 106, 147

cell phone(s), 1, 3, 5–6, 10–15, 17–18, 20, 24–25, 29, 31, 47, 49, 50–51, 54–55, 57, 66, 73, 75, 78–80, 83, 89, 91, 95, 98–99, 101–6, 109, 117, 122–23, 125–33, 135–44, 147, 149, 154, 163, 165

cellular service, 13

cellular system, 11, 103

channel(s), 101, 103–5, 147

charge(s), 38, 53, 55, 81–82, 110, 115–16, 120–21, 148, 151, 154, 157–58, 160

chemical reaction. *See* reaction(s): chemical

chemical(s), 18, 22, 63, 118, 148

chemistry, 114

chemists, 19, 53, 148

chip(s), 14, 31, 33, 46–47, 49–51, 56–59, 63, 65–66, 69, 73, 75, 79–80, 104, 128–29, 148, 152, 155–56, 159

chlorine, 26, 114, 116

circuit(s), 45–47, 49–51, 56–57, 65–66, 115, 117–19, 127, 146–48, 151, 155–57, 159

clean room, 58, 148, 152

cloud(s), electron, 112

code, 7–8, 78–79, 92, 94–95, 127, 147–49, 155, 158, 160, 162

code division multiple access, 106, 147, 149

coding, 100

communication, 5–7, 14, 91, 94, 99, 103–4, 132–33, 149, 152, 155, 159–60, 163

compact disk (CD), 34

compiler, 78–79, 147, 149

component(s), 14, 45–46, 53, 66, 127

computer programs, 45

computer(s), 31–37, 41–42, 44–47, 49–51, 57–58, 66, 74–80, 83, 88–89, 95, 100, 106, 127–32, 143, 146–50, 154–58, 160–64

computing power, 49, 58, 66, 129, 149

conductor(s), 51, 55–56, 95, 149, 162

contact(s), 17, 70, 114–16, 138, 151

copolymer(s), 25, 146, 149, 161; block,

26, 149; random, 26, 149

crosslink(ing), 26–29, 149, 154

crystal(s), 21, 51, 61, 64, 149, 154, 160

crystalline, 20–21, 51, 55, 59, 149, 162

current(s), 3, 97, 129

data line, 81–82

decision(s), 37–44, 56, 74, 80, 103, 127, 130

DEC PDP-1, 50, 166

decode(ing), 9, 14, 93, 95, 100, 127

density, 29

design, 13, 19, 69, 125–31

designer(s), 58, 126

digestion, 23

digital, 99–100, 146, 150, 152

discharge(ing), 81–82, 115, 117, 120–22, 150, 157

display, 31–36, 47, 49–50, 75, 83, 106, 109, 127, 131

elastomer(s), 28–29, 151

electric charge, 53, 61, 110, 148, 158, 164

electricity, 50, 55, 75, 95, 96, 109–10, 114–15, 117–19, 127, 136, 148–49, 151, 155, 158–59, 161

electrochemical cell, 110, 114–17, 146–47, 151, 153

electrode(s), 115–20, 122, 146–47, 151, 153

electrolyte(s), 115–16, 118, 120–22, 146–47, 151

electromagnetic radiation. *See* radiation: electromagnetic

electromagnetic spectrum, 98

electromagnetic waves. *See* wave(s): electromagnetic

electron(s), 52–57, 61–63, 65, 75, 80–82, 109–22, 138, 145–48, 150–51, 153–55, 158–64

element(s), 23, 26, 28, 33, 53, 56, 110–11, 113–15, 121, 138, 144–51, 153–56, 158–60, 162–63, 165

energy(ies), 24, 63, 98, 109–10, 112–16, 120, 130, 138, 151–53, 155, 160, 164

engineer(s), 19, 24, 27–28, 42, 44, 126, 130
entrepreneur(s), 131
error correction, 99, 152
experiments, 23

FAA, 137
FCC, 94, 143
FDMA, 106, 152
feature(s), 126, 128–29, 131, 133
Federal Aviation Administration, 137
Federal Communications Commission, 94, 143
field(s), 4, 19, 24, 54, 73, 95–97, 136, 148, 152; electric, 95, 96; magnetic, 95, 96
filter(s), 58, 93, 103, 152
fission, 112, 152, 158
flash, 75, 152
flash memory. See memory: flash
fluorine, 26, 114
force(s), 95–97, 152–53, 156, 164
frequency division multiple access, 105
frequency modulation, 99
frequency(ies), 94, 97–101, 103–4, 138, 140, 143, 147, 152, 154–55, 161
fusion, 112, 152–53, 158

gallium arsenide, 51, 146, 153
gas, 22, 24, 60, 63–64, 114, 120, 158, 160, 164
Gates, Bill, 39
global positioning system, 133
GPS, 133
gram, 54, 153
gravity, 54, 160, 164

half-reaction(s), 116–17, 119, 146, 147, 153
halogen(s), 114
hardware, 33, 44–47, 74, 76, 78–79, 122, 127–28, 154
helium, 53, 113, 153, 163
hertz, 98, 154
Hewlett, William, 132
Hiroshima, 138

honeycomb pattern(s), 103
hydrocarbon(s), 24–**27**, 147, 154
hydrogen, 22–24, 26–29, 53, 111, 114, 121, 145, 153–55, 158, 160; bonds, 28
hydroxide, 118–19, 121
Hz, 98, 154

IC(s), 50–51, 61
implant, 63, 68, 154
implantation, 62
ingot, 61, 150, 154, 164
injection molder, 24
input(s), 2, 37, 38, 41–42, 44, 80, 131, 156, 159
insertion sort. See sort: insertion
instruction(s), 33–36, 41, 44–47, 49, 56, 66, 74–81, 83–84, 89, 146, 148–49, 152, 155, 157–58, 160, 162
insulator(s), 18, 51, 55–56, 155, 162
integrated circuit(s), 50–51, 56–58, 66, 155
interference, 100, 136–37, 155
Internet, 73, 106, 129, 131–32, 164
interpreter, 78–79, 146, 155
iodine, 114
ion(s), 61–63, 68, 110, 115–18, 120–22, 145, 151, 154–55, 158, 160, 163–64
ionizing radiation. See radiation: ionizing
iron, 96, 121, 151, 156
isotope(s), 53, 110–11, 155

jam, 141, 155
jamming, 140
Japan, 138
Jupiter, 54

krypton, 113, 158

landline, 1, 9–10, 103, 140, 155, 160
language: assembly, 78–79, 146; high-level, 78; machine, 79
Lego(s), 45–47
light(s), 11, 14, 23, 34, 37–39, 44, 54, 63, 69, 94, 101–2, 104, 109, 127, 152, 154, 156, 159, 162–63; visible, 98, 151
light, speed of, 95, 97–98, 151, 162

linear, 26–27, 156
liquid, 22, 24, 56, 61, 63, 115, 151, 160
lithium, 120–22
lithium cobalt oxide, 120
lithium oxide, 120
lithium-ion (battery), 120
lithium-polymer (battery), 120
logic gate(s), 38–44, 50, 56, 74–75, 78, 80, 153, 156

manganese, 118–19, 156
manganese dioxide, 118–19
manganese oxide, 118
manufacture(ing), 18, 58, 68, 75, 127, 130–31, 146, 152, 156, 159
market research analyst, 131
mass, 54, 150, 153, 156, 164
material(s), 17–25, 28, 51, 55–57, 59, 60, 62–64, 67–69, 115, 120–22, 130, 141, 145, 150, 153–54, 156–57, 162–64
materials science, 19, 130
medical instrument(s), 137
memory, 28, 36, 66, 73–76, 80, 83–84, 88, 121–22, 129, 152, 154, 157, 159, 161, 164; flash, 75; persistent, 75, 152; volatile, 75, 161
merge sort. *See* sort: merge
metal(s), 3, 18, 29, 55, 59, 65, 68–70, 80, 114, 121–22, 145, 162
microcode, 78–80, 82–83
microphone, 139, 150
microscope, 23
Microsoft, 39
microwave(s), 98, 151
minicomputer(s), 49, 50, 66, 157
mixture, 22, 157, 162
mobile telephone switching office(s), 102
model(s), 1–4, 23, 27, 32, 45, 62, 64, 68, 78, 80, 83, 110, 126, 131, 156, 157
molecule(s), 19–23, 26, 28–29, 55, 59–60, 64, 97, 111–12, 115–16, 121, 148–49, 151, 154, 157, 160, 162, 164
monomers, 23–26, 28, 146–47, 149, 156, 158, 161
Morse code, 8

Morse, Samuel F. B., 8
MTSO(s), 102, 103

nanometers, 58–59, 148
neon, 113, 158
network, 12, 14, 76, 102, 129–30, 132, 147, 156, 157, 160
neutron(s), 53, 110–12, 155, 158
New Mexico, 2, 58, 168
Newton, Sir Isaac, 4
nickel, 121
nickel-metal-hydride (battery), 121
nitrogen, 26, 162
noble gas(es), 113, 158
non-rechargeable, 109
NOT gate, 38, 40–42
nuclear, 112, 114, 138, 152, 153, 158, 161
nuclear reaction. *See* reaction(s): nuclear
nucleus/nuclei, 110–12, 138, 151–53, 158–61

Olsen, Kenneth, 132
OR gate, 38, 41–42
orbit(ing), 33, 110–12, 154, 159
output, 42, 44, 156
oxygen, 22–23, 26, 53, 59, 64, 111, 159, 162

particles, 53, 55, 58, 110, 117–18, 148, 159
pattern, 21, 25–26, 31, 42–43, 47, 55, 61, 95, 96, 101, 103, 106, 145–46, 149, 159, 161
Pauling, Linus, 112
PC, 84, 88
periodic table, 53, 114, 144, 159
persistent memory. *See* memory: persistent
personal computer, 84, 167
phosphorus, 26, 56, 61–62, 68, 150, 159
photoresist, 63–65, 68, 156, 159
physicists, 19, 53, 159
pixel(s), 33–35, 47, 49, 159
plastic(s), 14, 17–20, 22–29, 46, 55, 65, 127, 130, 144, 149, 160
plutonium, 138

polymer(s), 19, 22–25, 27–28, 120, 146–47, 149, 151, 154, 156, 159–61; beads, 24

polystyrene, 26, 160

potassium, 114, 116, 118–19, 160

potassium hydroxide, 114, 118

pressure, 24, 52, 148, 164

privacy, 139

processing, 19, 24, 31, 46–47, 57, 61, 63, 65–66, 83, 130, 144, 149–50, 156, 161

program(s), 37, 44, 47, 50, 78–79, 132, 146, 148–50, 155, 160, 164

programmer, 44, 50, 78–79, 146, 154, 158, 162

programming, 44, 73, 78–79, 84, 146–47, 154, 156

project manager, 131

property(ies), 17–19, 22, 25–26, 51, 55, 59, 110, 112, 121, 130, 146, 150, 156–57, 159–60, 163

protocol(s), 92, 94–95, 100, 103–4, 106, 128–29, 149, 154, 160, 163

proton(s), 53, 61, 110–12, 115–16, 154, 158–60

PSTN(s), 103, 160

public switched telephone network(s), 103

radiation, 98, 138, 155; electromagnetic, 138; ionizing, 138

radio, 12, 95, 98–103, 109, 129, 133, 136–41, 147, 151, 155, 161

radio signal. See signal(s): radio

radio station(s), 94, 136

radio waves. See wave(s): radio

radon, 113, 138

RAM, 75, 80, 81, 83, 161, 167

random access memory. See RAM

random copolymer. See copolymer: random

reactant(s), 115–17, 120

reaction(s), 64, 113–16, 119–21, 148, 158, 161; chemical, 23, 25, 59, 109–11, 113–16, 120–21, 148, 151, 153, 158–59, 161; nuclear, 112

read-only memory. See ROM

rechargeable, 109, 120–22, 158, 161

refine(ing), 25, 119

reuse, 36

Rio Rancho, 58

risk(s), 131, 137, 139

ROM, 75, 80–81, 161

rubber, 28, 56, 122, 164

sample(ing), 100, 146

sand, 59, 64

science, 3, 4, 11, 15, 19–20, 54, 125, 130–31, 142, 144, 148, 157, 161–62, 168

scientific method, 19, 161

scientist(s), 19–24, 27, 42, 130–31, 146, 148, 156, 160, 162

semiconductor(s), 14, 31, 33, 46, 50–52, 56, 60–61, 63, 66, 68, 128–30, 144, 146, 148, 150, 152–54, 156, 159, 162, 164; chips, 129, 130

shell(s), 17–19, 23–25, 29, 46, 111–14, 116, 121–22, 127, 145, 151, 153, 162–64

signal(s), 7, 9, 12, 14, 39–40, 75, 92–95, 99–106, 109, 127, 133, 136, 138–41, 145, 147, 149, 152, 155, 161–63; radio, 95

silica, 59, 64, 69, 162

silicon, 26, 51, 56–57, 59–64, 68–69, 146, 148, 150, 153–54, 157, 160, 162

silicon dioxide, 59, 64

software, 14, 44–45, 47, 66, 73–79, 81, 83–85, 89, 104, 106, 122, 127–30, 139, 143–44, 154, 162

solid, 20, 22, 42, 150–51, 153–54, 160, 162

solution(s), 4, 44–45, 115, 118, 146–47, 157, 162

sort: insertion, 85, 88; merge, 85–86, 88

source code, 78–80, 83, 146, 149, 155, 158, 162

strength, 27–28

subparticle(s), 110

substance(s), 22, 29, 53, 63, 115, 121, 145, 147–51, 154, 156–57, 160–64

sugar, 22–23, 53

supercomputer, 84, 88

switch, 37–39, 50, 56, 74, 75, 80, 82–83, 163

TDMA, 105, 163
telegraph, 8–9, 163; key, 8; sounder, 8
telephone(s), 9, 94–95, 147, 157–58, 160
temperature, 24, 148
text message, 32, 35, 47, 73, 75, 83, 106; messaging, 8
time, 2, 7–9, 13–14, 24, 27, 29, 36, 42, 45, 49–51, 62–63, 65–66, 74, 76, 79, 82, 92–94, 97–98, 100, 103–5, 109, 128, 132, 135, 137–38, 155, 160–61, 163
time division multiple access, 105, 163
tone(s), 95
tower(s), 10–12, 102–5, 128–29, 135–38, 141, 147, 149, 152, 154, 156, 157, 163
transistor: gate, 56
transistor(s), 46–47, 49–51, 54, 56–67, 69–71, 73–75, 78–83, 152–53, 159, 161–63
Turing, Alan, 44–45, 166

United States, 60, 94, 98, 112, 140–41
uranium, 138

vacuum, 50, 55, 97, 151, 164
vacuum tube(s), 50, 164

valence, 111–15, 121, 145, 148, 153, 163, 164
visible light. *See* light(s): visible
voice, 9, 47, 129
volatile memory. *See* memory: volatile
voltage(s), 51–52, 56–57, 82, 127, 146–47, 162, 164
Von Neumann, John, 44–45
vulcanization, 28, 164

wafer(s), 51, 59, 61–65, 68–71, 148, 150, 154, 156, 159, 164, 166
walkie-talkie, 12
water, 17–18, 22, 24–25, 80, 96–97, 114, 116, 118, 121, 151, 153
wave(s), 95–102, 136–39, 145–47, 149–52, 154–55, 157, 161–64; electromagnetic, 95–98; radio, 95–96, 98, 101, 136
wavelength(s), 63, 97–99, 164
weight, 18, 54, 123, 138, 156, 164

zinc, 118–19, 165
zinc oxide, 118–19